APPLIED MATHEMATICS

Applying Modeling to Everyday Life

Erik Richardson

Cavendish Square
New York

Published in 2017 by Cavendish Square Publishing, LLC
243 5th Avenue, Suite 136, New York, NY 10016

First Edition

CPSIA Compliance Information: Batch #CS16CSQ

All websites were available and accurate when this book was sent to press.

Library of Congress Cataloging-in-Publication Data

Names: Richardson, Erik.
Title: Applying modeling to everyday life / Erik Richardson.
Description: New York : Cavendish Square Publishing, [2017] |
Series: Applied mathematics | Includes bibliographical references and index.
Identifiers: LCCN 2016006886 (print) | LCCN 2016016490 (ebook) |
ISBN 9781502619693 (library bound) | ISBN 9781502619709 (ebook)
Subjects: LCSH: Mathematical models--Social aspects--Popular works.
Classification: LCC QA401 .R485 2017 (print) | LCC QA401 (ebook) |
DDC 511/.8--dc23
LC record available at https://lccn.loc.gov/2016006886

Editorial Director: David McNamara
Editor: B.J. Best
Copy Editor: Nathan Heidelberger
Art Director: Jeffrey Talbot
Senior Designer: Amy Greenan
Production Assistant: Karol Szymczuk
Photo Researcher: J8 Media

The photographs in this book are used by permission and through the courtesy of: Roberto Machado Noa/LightRocket via Getty Images, cover; tovovan/Shutterstock.com, 4; duncan1890/E+/Getty Images, 6-7; PBNJ Productions/Blend Images/Getty Images, 10; belterz/E+/Getty Images, 12; André Hatala/File:Frans Hals - Portret van René Descartes.jpg/Wikimedia Commons, 15; William Playfair/File:Playfair TimeSeries-2.png/Wikimedia Commons, 16; Florence Nightingale/File:Nightingale-mortality.jpg/Wikimedia Commons, 17; Charles Joseph Minard/File:Minard-carte-viande-1858.png/Wikimedia Commons, 18; LANL/File:JohnvonNeumann-LosAlamos.gif/Wikimedia Commons, 20; angelhell/E+/Getty Images, 28; Dusit/Shutterstock.com, 33; marekuliasz/Shutterstock.com, 41; Master3D/Shutterstock.com, 43; stockyimages/Shutterstock.com, 46; Jupiterimages/Stockbyte/Thinkstock, 49; Mike Pont/WireImage/Getty Images, 52; Robert Kyllo/Shutterstock.com, 54; alengo/E+/Getty Images, 58; amateur with luck/File:Kesagami whitewater canoeing C-2.jpg/Wikimedia Commons, 67; Spc. Nathan GoodallFile:U.S. Army Spc. Travis Kerr, right, an intelligence analyst with; Headquarters and Headquarters Battery, 17th Fires Brigade, gives a briefing to; the brigade staff during an exercise at the Mission Command 130605-A-AU369-; 870.jpg/Wikimedia Commons, 70; Visuals Unlimited, Inc./Carol & Mike Werner/Getty Images, 78-79; autsawin uttisin/Shutterstock.com, 83; KDS4444/File:Crystal ball by Ron Bodoh.JPG/Wikimedia Commons, 84; Qc72/File:Population pyramid matelica 2010.png/Wikimedia Commons, 86; Ploty, 87; pmphoto/Shutterstock.com, 91; Alain r/File:FY221c15.png/Wikimedia Commons, 96; RoshiniA/File:Flow regimes in a street canyon.png/Wikimedia Commons, 99; Paul venter/File:Francis guthrie.jpg/Wikimedia Commons, 100; Fibonacci/File:World map with four colours.svg/Wikimedia Commons, 100; Robert Adrian Hillman/Shutterstock.com, 103; Bryan Chan-Pool/Getty Images, 106-107; Cromagnon/Shutterstock.com, 109; Peter Fitzgerald/File:Chicago violent crime map.png/Wikimedia Commons, 110; Minerva Studio/Shutterstock.com, 113; Image Source/Getty Images, 116-117.

Printed in the United States of America

TABLE OF CONTENTS

A map can reduce a complex city into a simplified model, making it easier to use.

INTRODUCTION

This book is going to take a walk through something like a museum exhibit of mathematical models. At each display along the way, you'll be able to stop and put on metaphorical headphones to hear the guide talk about the way each model is used, what sets it apart from other models in the exhibit, and how different jobs and activities have certain kinds of models that are central to their successes.

Before we start the first chapter, which gives an overview of some of the key historical figures in models, it would serve us well to spend a few minutes talking about what we actually mean by the term "model."

What Is a Model?

Reflect for a moment about what it would mean to be an exact map of the world. In order to make a map that was a representation of the world in every detail, leaving nothing out, it would have to be as big as the world itself. As soon as we started to shrink it down, some details would be lost. This stone would be gone, or that tree, or the patch of grassy field just over there. If we shrink the scale, some things would no longer be visible, even if they were somehow still there in theory. They would no longer be part of the map we are seeing and using to navigate.

At the same time that a map cannot include everything, if we reduce it down too far, it no longer contains enough detail to be helpful to us. A map of

NORTH POLE
GREENLAND
NORTH AMERICA
NORTH PACIFIC OCEAN
NORTH ATLANTIC OCEAN
Tropic of Cancer
GULF OF MEXICO
CARIBBEAN SEA
West Longitude
THE EQUATOR
SOUTH AMERICA
SOUTH PACIFIC OCEAN
Tropic of Capricorn
Antarctic Circle
ANTARCTIC OCEAN
SOUTH POLE

Globes and world maps, like this one from 1837, allow us to see the whole world by leaving out lots of details and information.

the city need not include every tree and shrub, but it must surely include all the streets, right?

Sometimes it could contain realistic pictures to help add interest and artistic value to the map, but it could also contain a wide range of different symbolic images—maybe a small triangle for trees, or simple squares to represent complex, **ornate** buildings. A map is a model of a city. It helps reduce something very complex into something we can more easily understand. As we go through the museum exhibit of this book, we will see a number of models that fall along that spectrum from too realistic to too symbolic.

What Good Are Models?

Why do we go to the trouble of constructing models—from maps to graphs to symbolic pictures? We can address the creation of a given model by connecting it to one or more of the following goals:

1. To provide a version of something that is smaller in scope so we can use it to guide our decisions in a way that connects back to the real world.

2. To bring certain features of the information concerned into clearer relief by taking away noise and clutter.

3. To add some dimension of sensibility to the numbers and concepts involved. By "sensibility" we mean literally the ability to be capable of being engaged with our senses.

4. To summarize and communicate some set of information, including both the **data** and the relationships between data points.

When we turn to consider the different ways those things can be accomplished, we can likely agree at the outset on a variety of ingredients for a model. We can use numbers, words, images, and variations in color and size in a diagram or model.

This engagement of the senses is especially true of sight. The majority of our brain is geared to the processing and storage of visual information about the world around us, translating concepts or changes or patterns into a visually accessible format.

It is also fitting to briefly address what is not being considered as a model. It is true that math equations and formulas are highly symbolic models. It makes perfect sense for those models to have been stripped all the way down to symbols to increase the speed and ease of running the models to test the outcomes. However, that speed can often come at the expense of a loss of clarity in understanding and applying what those quickly calculated numbers tell us about things back in the real world.

To add an engaging bit of a twist as we head in, consider that letters and words and sentences suffer the same limits and fate as other symbols—they have left out some detail so as to be small enough to be easily used. Hopefully you will appreciate, by the time you are finished with your tour, that this introduction stood no chance at being a complete and rigorous definition of "models."

Models are not just about physical objects. They can also be used to represent things like time and the seasons, as on this ancient Aztec calendar.

ONE

The History of Modeling

Like the problems with maps that we just discussed, I'm not sure how to give this chapter a title that is both meaningful and much shorter than the chapter itself. I can't really say it's a true "history" of models, because we will not even come close to covering the whole history. I also can't just say it's a history of "models," because we will only consider a very small selection of all the possible models that could be covered.

I will leave it to you to decide what you think would make a good title for the chapter after you've read it and thought about it. At least you will now be going in with appropriate cautions about the scale or significance of the chapter. You will also, I think, have a little bit of appreciation for how words and names and titles fall into some of the same possible problems that I mentioned in the introduction.

Not All the History, Not All the Models

Why not include all the history? Well, two reasons: 1) The whole history would take us all the time since cavemen started drawing with sticks, and half a history would still

Sometimes models help show how two or more things are related. Here we see the relation between natural features and the boundaries of nations.

Descartes's many talents

Rene Descartes didn't just wander along through life waiting for his big break. He didn't happen to get lucky and write a couple historic books that would change the way people thought about everything from math to politics.

Descartes was born in France in 1596 and was homeschooled until he was eight. He then went off to a Jesuit school, where he graduated at eighteen. Jesuit schools—the best private schools around—were rigorous, and Descartes was outstanding. Math was his strongest area.

He headed off to Paris, where he graduated in 1616 with a law degree, and then he enrolled in a military school. In 1619, at the age of twenty-three, he signed up with the Bavarian army and was a soldier for nine years. When he ended up in Holland, he left the military to pursue a scholar's life in math and philosophy.

Along the way, he also found time to discredit some con men claiming mystical powers, study medicine and become a doctor who provided free care, study engineering, write a book about music, advocate for human rights and religious freedom, write back and forth with the princess of Bohemia developing a theory of morality, and serve as tutor to the Queen of Sweden.

take half of all the time since cavemen started drawing with sticks. 2) Our models have gotten much more complex than simply drawing with sticks.

Why not include all the models? Here, we can pin it down to one reason: There are countless possible **pictograms** and diagrams and plates of symbol salad that could be talked about—from Mayan calendars to computer **turbulence** models. Spending time on each and all of them would only add small amounts to our understanding of different approaches. Instead, I just want us to focus here on the small pile that are the basics of models that would account for probably close to 90 percent of all the models that more than 90 percent of the people reading this book would ever need (or want!). Of course, I may only be 90 percent accurate in my estimates (including this one).

Graphs and Diagrams

Descartes reinvents mathematics

In 1637, Rene Descartes published a phenomenal piece of writing titled *Discourse on Method*. Perhaps an even greater example of his genius is the fact that he wrote the *Discourse* in French (rather than Latin) so, as he said, everyone of good sense—including women—could read it and learn to think for themselves!

The *Discourse* was a massive work that attempted to rebuild all our thinking—including science, philosophy, and politics—on the clear, sturdy framework of mathematical reasoning. Along with his reconstruction of much of the knowledge of his time on this new framework, the book also contained three essays at the end involving pursuit of truth in particular areas of science. In "Dioptrics," he developed the law of **refraction**; in "Meteorology," he explained how rainbows work, and in "Geometry," he laid out his revolutionary analytic geometry, which allowed for mapping algebraic **functions** onto a visual set of coordinates. In the process, he also improved the system of using letters for variables as

Rene Descartes was astonishing. He revolutionized science, invented a whole new approach to mathematics, and rebuilt philosophy from the ground up.

well as the superscript form for exponents, both of which made mathematical calculations easier and clearer.

Playfair and others

In 1786 a Scottish economist and engineer named William Playfair published a book called *The Commercial and Political Atlas.* This work attempted to represent the whole of English **economics** of the era. It seems to have been the first to use such now-common tools as line graphs, bar charts, area charts, and histograms. A later work, called *Statistical Breviary* (published in 1801), is also credited as the first to introduce the pie chart. These works, the *Atlas* in particular, were considered great examples in demonstrating

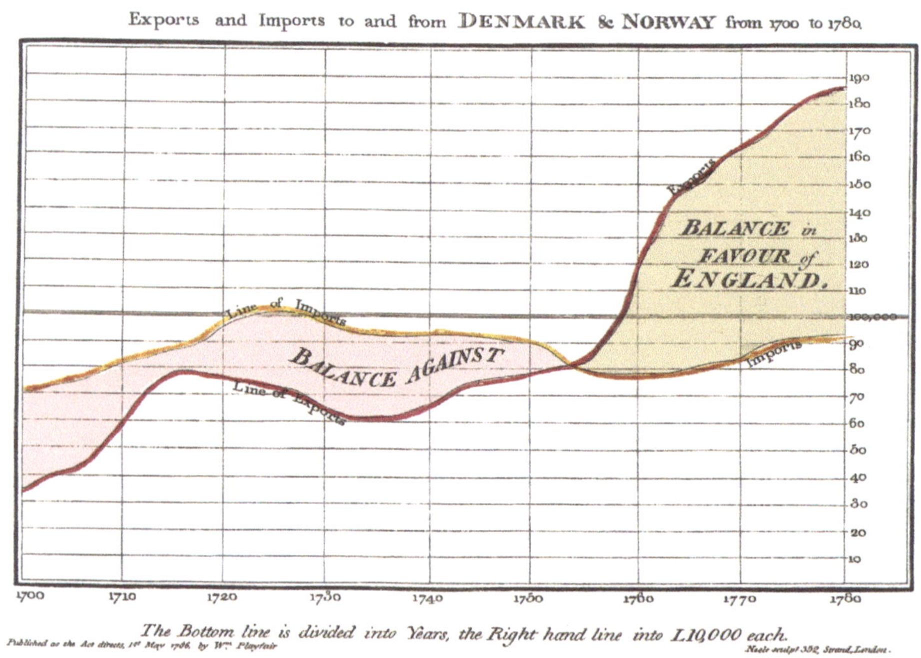

This is an example of a graph from William Playfair's famous book. It shows exports and imports to and from Denmark and Norway based on his data collection.

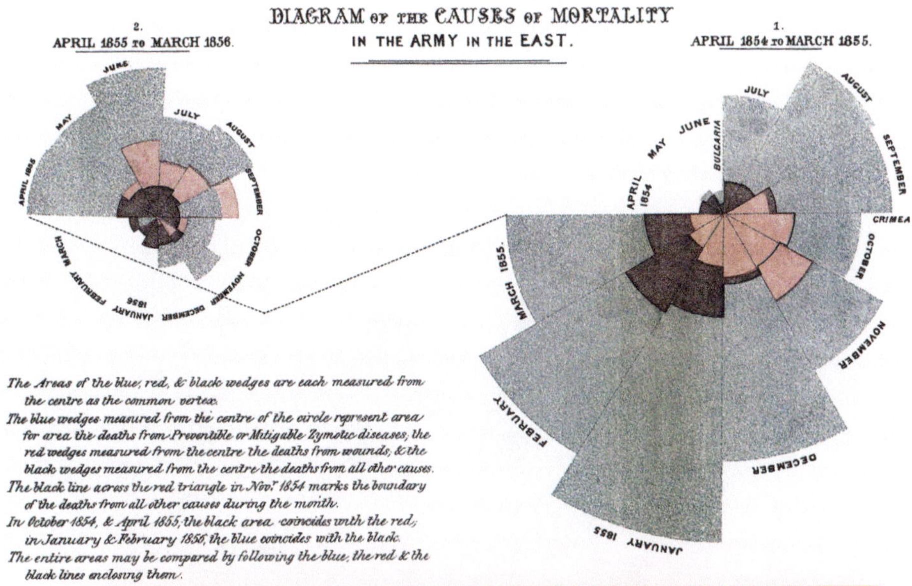

This graph shows an example of Florence Nightingale's pioneering work in the use of statistics to help improve soldier mortality and hospital practices throughout Europe.

how the graphing of data could help reveal patterns and open up potential lines of research.

The polar area diagram was first presented by André-Michel Guerry. He used one in 1829 to represent daily and seasonal variations in wind direction. Building on this, Léon Lalanne used polar diagrams in 1843 to show **frequencies** of wind direction arranged around the point of the compass—a tool still used by modern meteorologists.

We see graphic models taken and applied to health care by John Snow and Florence Nightingale. Snow plotted out the cholera deaths in London in 1854 to try to trace back to the original point of outbreak for the disease. The case of Nightingale, the English nurse, is somewhat more famous. Her model combined ideas of polar area and stacked bar graphs to present the number

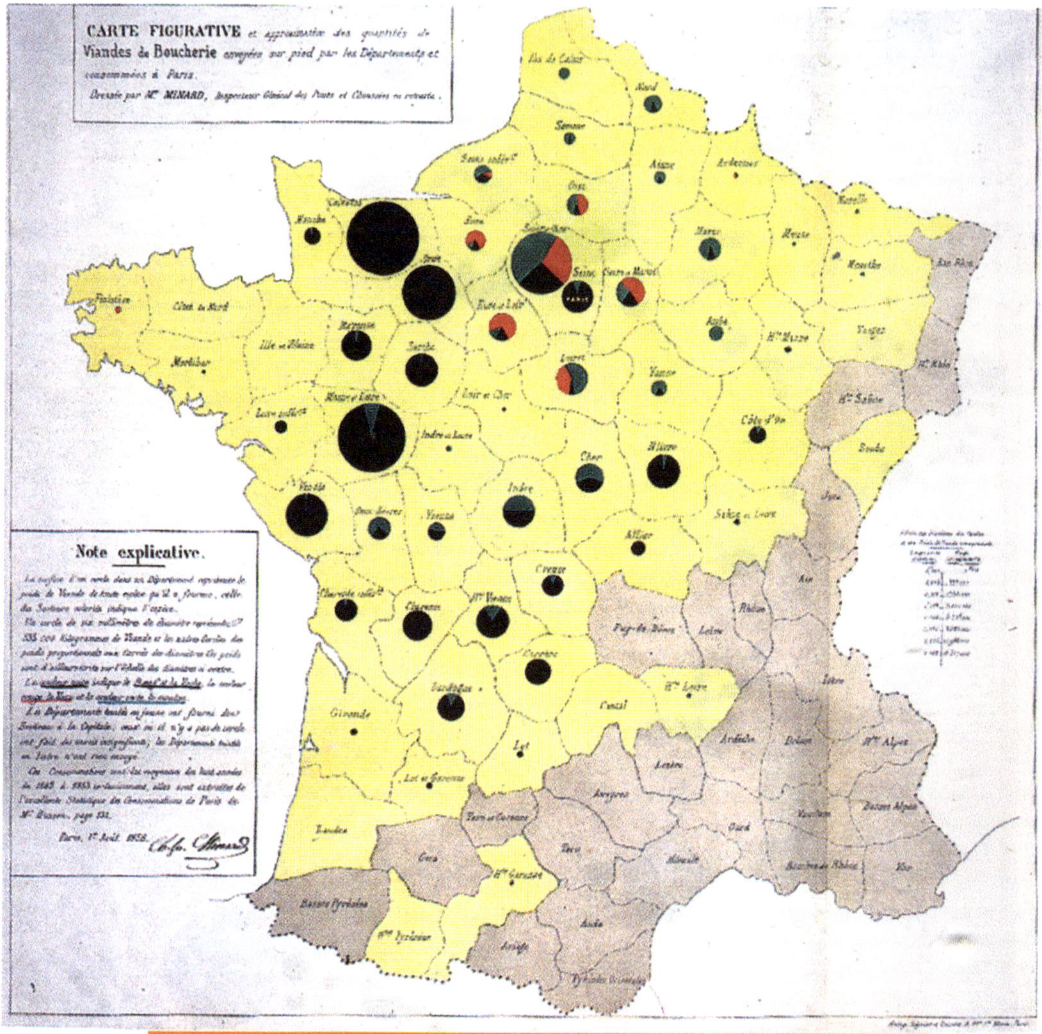

In this map, one of the earliest uses of pie charts shows the amount of cattle shipped from different areas in France.

of deaths from different causes for soldiers in the Crimean War. She also leveraged the use of such graphical tools to convince Queen Victoria to support improvements to the conditions and practices in military hospitals.

In 1858 we see the first real use of pie charts, as we think of them, by Charles Joseph Minard, a French engineer. Among his models were pie charts showing the cattle sent to feed Paris from different areas of France. He is also know for a large **portfolio** of maps, including one showing Napoleon's disastrous Russian campaign. Of particular coolness was how his model was able to show four different variables in a two-dimensional diagram.

The last item to cover in this stage is the 1878 set of diagrams by *Nature* magazine. In them, James Joseph Sylvester created a set of diagrams showing how mathematical properties and chemical bonding are related.

History of Game Theory

Game theory is the mathematical field that studies **strategy** options in situations where the costs and benefits for each participant depend on the decisions of the other participant(s). Game theory uses math to figure out your best strategy when the results depend on the other player's strategy (or strategies if there are more than two players).

Precursors of game theory

In looking back, there are some early discussions of two-person games. Three in particular stand out as precursors to modern mathematical game theory. The first is a letter written in 1713 by Charles Waldegrave to his uncle, a British diplomat. In it, he offers an analysis of a particular card game called "le her." A second interesting example comes from an analysis by James Madison of different possible responses of states under different possible systems of taxation.

John von Neumann was a mathematical genius whose ideas and research impacted almost every field of modern science and computing.

John von Neumann's intelligences

John von Neumann, the Hungarian mathematician, was born in Budapest in 1903 and died in Washington, DC, in 1957. He was a genius who, as a child, could tell jokes in Greek and could recite memorized pages from a phone book. He went on to impact almost every area of mathematics, but one of his greatest roles was being the first mathematician to make breakthroughs in using math in the **social sciences**.

After studying in Budapest and Zurich, von Neumann was invited to study and teach at Princeton University. In 1933, he joined the Institute for Advanced Study in Princeton, and he taught and researched there for the rest of his life. His work in different fields resulted in several honorary doctorates, academy memberships, prizes, medals, and other distinctions.

Von Neumann was interested in a lot of different subjects, and he made important contributions to set theory and logic. He created a kind of geometry that didn't have "points." Von Neumann's work in physics and in quantum mechanics helped create several breakthroughs.

On top of that, he was endlessly interested in the relation between math and machines. He established the logic used in computer design, and some of his ideas about artificial intelligence are still being applied to research.

In a work titled *Researches into the Mathematical Principles of the Theory of Wealth* in 1838, Antoine-Augustin Cournot analyzed the situation of two firms controlling a market for some product, and his solution is a small example of what would later be known as a Nash **equilibrium**. In a logical next step, we see someone applying game theory analysis to the game of chess with an article by Ernst Zermelo in 1913. He opened the door to broader ways of thinking about game theory through his treatment of the optimal chess strategy.

Game theory comes into its own

It was the publication of a paper in 1928 by the math giant John von Neumann that gave birth to game theory as a properly recognized subfield. He developed his ideas over the next few years, engaging on the new topic with great thinkers of his day, resulting in a book in 1944 called *Theory of Games and Economic Behavior*. Some of the strategies developed in the text help to show how to find the best possible solutions to certain kinds of games. These games and solutions were based on situations where the two players could be convinced to cooperate based on various possible forms of agreement and consequences for breaking the agreements.

The first discussion of a game type where the players are not in position to form an agreement or set out conditions of enforcement was by Melvin Dresher and Merrill Flood in 1950. They were two mathematicians working with RAND Corporation to develop applications of game theory for military strategy in the new context of nuclear weapons.

In the years since, John Nash and four other Nobel laureates, along with numerous other mathematicians, political scientists, and economists, have gone on to develop what has become one of the most important tools for modeling and analyzing complex, real-world puzzles in fields from evolutionary biology to law enforcement to finance.

While the approach in the coming chapters attempts to focus on the different varieties of graphs and charts rather than game theory itself, it is important not to be misled about the **comparative** scale or importance of the two different areas. Graphs help you see what has occurred. Game theory helps you solve a situation where a whole tangle of line graphs and functions are in motion at the same time.

As you head into the next chapter, let me send you off with a bit of a game theory mindset. How will you choose and how will you benefit? You can choose to be engaged and curious *or* just kind of wandering along. You can view the text as having interesting and useful ideas *or* just kind of wandering around. Think about the different outcomes depending on which strategy you put into play, even if you don't know which strategy the text has picked.

TWO

Modeling in Your Everyday Life

In this chapter, I want us to wander through different possible areas and activities that are part of our day-to-day routines. The items might not all apply to you, but hopefully the kind of activity will resonate with some of the other things you do. You might not do tae kwon do, for example, but you might participate in some form of organized training or sport. More important than the exactness of the activities chosen, though, is the process of showing how aspects of that activity help us determine what kind of data would be useful and what kind of models and graphs would add some useful insight. The information could help us make choices on how to spend our afternoon or make decisions about whether to keep spending money to repair our car instead of buying a new one.

I also want to help you engage with the ideas and questions because the process of thinking "about" math uses some of the same skills and provokes some of the same curiosity as using math to think "about" other things. Using equations to solve a problem is mathematical thinking, but so is figuring out which formula or equation you need. Understanding what the tools do and which ones are best plays into the bigger picture of using mathematics effectively.

To help with that process, each area will talk about what tools would be helpful in getting a good, meaningful understanding of the data. In addition to that, I've also tried to help spark your thinking about other areas that would be likely applications for the same kind of model.

Of course, the other thing is to remember to let yourself have at least a little bit of fun thinking about some of the different concepts or ideas you happen to bump into along the way.

Fitness and Nutrition

As many professional coaches, athletes, and trainers will tell you, an important part of successful fitness programs—whether you are training for the Olympics or just training to improve your overall health—is being able to adjust your training based on the data. It's helpful to know what is working or not working at more than a **vague** level where it "seems like" you're getting stronger or recovering faster. If you have that data about yourself, you can use it to compare yourself, your personal bests, and your goals to the numbers of other successful athletes. That can help you pace yourself so you're not overtraining or expecting unreasonably fast results. It can also help you take advantage of different research materials that apply to your fitness level.

It's interesting to develop various tools for measuring and thinking about the world around us and then turn those tools back on yourself to see what you can learn. It's a bit like trying to look at yourself through a microscope. In this case, let's consider the examples where a column chart with two sets of data would help to quickly grasp key comparisons for our training. For example, this kind of graph allows you to compare your daily caloric intakes to some of the top runners in the world.

Meet Chris Friesen

Please meet Chris Friesen, now in his fifth year as a trainer at Average Joe's Fitness in Manitoba, Canada.

So, tell us: What are some different kinds of math that you use in your job?

I'm a huge believer in tracking training data, because you can't change what you don't track. I often ask clients specific questions about things that they did last week, last month, or even last year to emphasize that memories are not as good as we think they are. Without a record of your past training, how can you make good adjustments to improve?

How would you say analyzing that data you've collected helps your clients make smarter decisions?

Even more, it's about developing habits that enable them to analyze that collected data for themselves and to educate them toward being their own trainers. In essence, you are educating yourself out of a job.

Do you think our own competitive nature kicks in once there are numbers that look like a score to beat?

Yes, seeing your numbers go up is a key factor. I've had clients who were grandmothers in their fifties who never lifted weights before getting really excited to push their dead-lift number up to 150 percent of their bodyweight. Each time they didn't look or feel different than the week before, but that number changed! There is just something about seeing people challenge their supposed limits, and tracking numbers is a huge part of making that happen.

The empowerment of showing someone how strong they can become translates to more than the physical, I think. The numbers are only a tool to help them see they can take control over those things and make a difference.

Any additional thoughts on how math, models, and graphs help you succeed in running the business or in helping you convince people the benefits of becoming a member?

I've been in the customer service industry for the last twenty-five years, and if there is one thing I have learned, it's that power comes from how you make people feel about themselves (whatever that looks like in any given situation), and how you can help people realize their goals, in my opinion.

Love the way you think about numbers helping people. Thank you for talking with us.

No, thank you for the opportunity to share my passion with you and the readers.

Sometimes making a graph or chart in 3-D allows us to not only show change over time but to compare and contrast two sets of data more easily.

One of the most common models to use for representing data is the bar chart. While we sometimes divide charts into bar charts if the data is read sideways and column charts if it is oriented vertically, they function the same.

In each case, they are being used to show the frequency of each category. The taller a given bar, the larger the number of items from the group that fall into that category. A primary benefit to this kind of display is that it allows for an easy comparison of relative amounts across the categories. To assist in that, we usually make sure to pick a scale that makes that process easier. Having a scale too large or too small will make the differences between the categories seem either smaller than they are, or much bigger. A requirement for being able to use these charts is that the items must be able to fit into discrete groups. (There is no overlap or "in-between" option in a bar chart.)

A regular column chart—for example, one that graphs your **aerobic** endurance—can be in 3-D, but the third dimension doesn't really add anything. But a version which uses that extra dimension of depth to add something like the national averages for aerobic endurance at your age level can be much more informative.

When you have put the data together, you are in a position to think about what it's telling you. Maybe your daily caloric intake is really high compared to others at your skill level, even though your strength gains have slowed. That could suggest the need to adjust the contents of your meals to healthier stuff to boost the ratio of protein to calories.

In addition to the use of bar/column charts, there are a few other key kinds of information modeling that are commonly used in helping to more clearly understand different aspects of our fitness and nutrition. The one you might think of first is the good old line graph. That allows us to show progress of a given variable over time—how your single-lift maximum weights have increased over time, or how your body fat percentage or running times have decreased. Another very useful tool is the pie chart, which allows you to more easily evaluate the balance between factors than a bar chart would. A good example of this would be to think about the quick visual measurement provided by a pie chart showing what percentage of your daily calories are coming from protein, from fat, or from carbohydrates.

As your training becomes more advanced, your models would also need to include some graphs of curved functions that help to give you a fair picture of how your progress and gains (or losses) will naturally diminish over time. That way you are not expecting (unrealistically) to make the same improvements later on as quickly as you did at the beginning of your training.

Besides fitness and nutrition, what are some similar areas or activities that would fall into the same sphere so that this kind of model would help us understand them and be more successful? A couple of examples that come to mind would be fantasy football or baseball and other kinds of ongoing individual activities you like to participate in, like rock climbing or archery, or maybe camping and backpacking. Each of those areas would utilize similar charts to compare your performance or your equipment to the performance or recommendations of other people who do the same activity. How does your pack weight compare to that of experienced hikers, for instance? Similarly what does your climb rating look like? What are your strengths and weaknesses compared to other climbers? You get the idea.

Pet Adoptions

Let's move beyond that **scenario** where we were tracking and comparing different pieces of data and comparing them to other groups. Here we should recognize the different scenarios where we might instead be comparing totals. But we might also need to see how separate ingredients each contribute to the changed totals.

We can also use bar graphs for a more complex data comparison by clustering or by stacking the different groups. For a grouped (or clustered) bar chart, we would set the different series right next to each other for a given category. An example of this might be when a business owner with more than one store wants to see each store's sales for January, then each store's sales for February, and so on. This is not quite as easy a tool for seeing just Store A's performance, comparing one month to the next, but in exchange, it makes it easier to compare Store A versus Store B versus Store C each time.

The other way to help add meaningful context to the data being modeled is to stack the different items on top of each other. By combining the different elements into one column for each category or time period, the graph allows us to more easily compare total outcomes between one period and the next, and it prevents a certain amount of a cluttered feeling in the graph, allowing us to potentially consider more time periods or more categories.

A good example to illustrate this would be a chart like the one below representing the numbers of each kind of animal adopted from the metropolitan animal shelter each month, with each color representing a different category—dogs, cats, birds, etc.

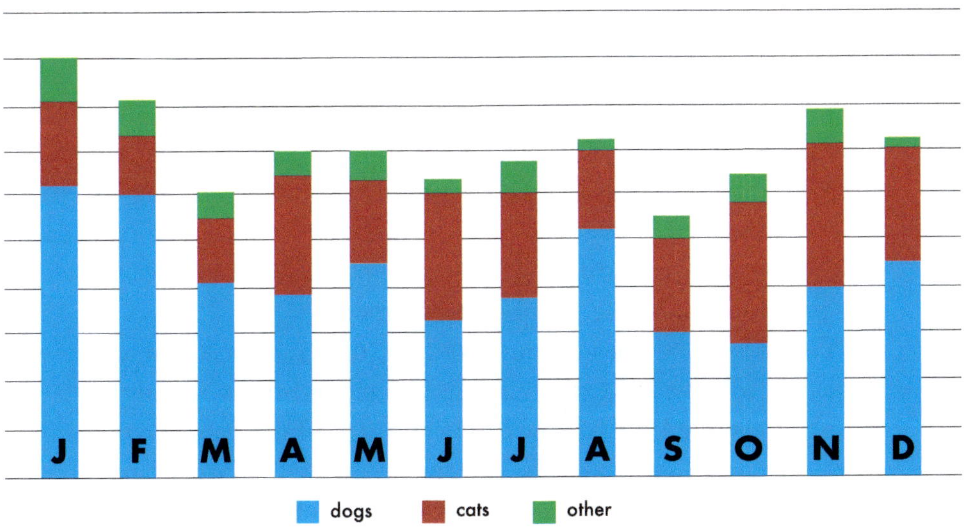

Reflecting on the example of the animal shelter, what do you think would be an example of a graph that would show a problem or weakness worth addressing? Maybe the animal shelter has overall totals that look the same, but with the stacked graph, they see that the number of dogs coming through for adoption is slowly climbing while other animals are showing up less and less. This could have an impact on having enough rooms for the growing dog population at the shelter. Dogs typically need more room to run around and play. The sooner the directors are able to see that transition start happening, the sooner they could start a fundraising campaign to help cover the costs of adding space.

Some of the other key tools that would be helpful in the operation of an animal shelter would be line graphs to track some of the straightforward numbers, like total contributions or total budget costs from month to month. Interestingly enough, another area where the use of a stacked graph like the one above would be particularly helpful is to track the different sources of donations. Fundraising is a huge function for many animal shelters, so they want the best, most detailed information about how their campaigns are going in terms of reaching different audiences.

In contrast, the kind of tools that would come in handy for tracking and analyzing the different costs, including medical procedures, would be a polar area graph. I'll talk about that a bit later, but it would visually give a sense of both the percentage of the total represented by each procedure and, since different medical procedures have very different costs, the overall cost impact of each category relative to the number performed.

Pet adoptions are just one example of a scenario where these tools would give us extra clarity. Consider how you might use exactly the same kind of chart to keep track of the money you earn from different jobs you help out with—mowing yards, part-time work at the fast-food restaurant, being the referee or umpire for a couple kids' sports on the weekends. This kind of stacked bar graph would allow you to see and compare how much you are earning from each of those activities and would still show the difference from month to month of the different totals you earned from all activities/jobs combined.

Going to the Movies

Let's say you decide to head out to a movie, even though it seems like the price has been going up. Maybe you are even wondering, on the way to the movie or while standing in line for a ticket, about why that keeps happening.

A great number of the graphs in mathematical models are representing functions, which result not in straight or jagged lines, but smooth, **continuous** curves. In understanding and representing the dynamics that go into the process of finding the right price of any given product, what is really at play is the interaction between two different curves or functions. One represents the willingness

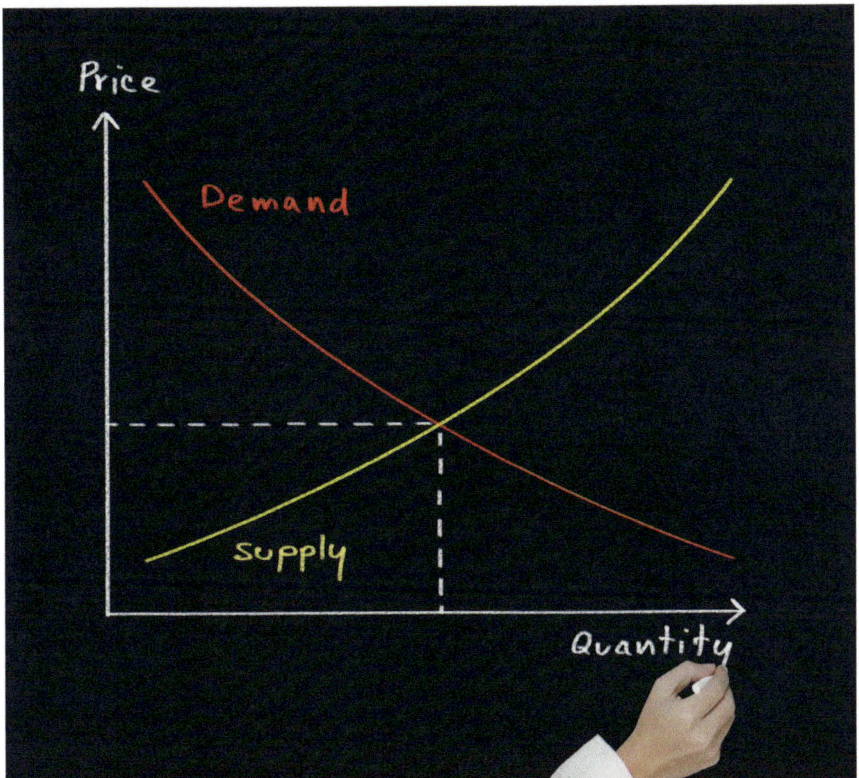

This standard graph is familiar to every student of economics, and it shows how supply and demand work together to establish the price of something.

of buyers to purchase items at various prices. Consider how you would be interested in buying fewer boxes of cookies if they cost $50 per box rather than $5 per box. This is called the demand curve. The other represents the willingness of sellers to produce and sell items at various prices. Consider how differently you would feel about working an extra hour this week if you were going to make $25 for the hour of work compared to if you were going to make $5. This curve is called the supply curve. The point where these two intersect is what becomes the settled price for an item—like the price of a ticket.

When you finally get through the ticket counter and then slowly inch your way to the front of the line at the candy counter, the price of every item has come about as a result of the interaction of demand and supply curves, just as the cost of the movie ticket had.

Seeing where the equilibrium price happens, though, doesn't actually give us an understanding of why it has been going up, nor does it simply allow the movie theater owner to adjust prices. Being able to use the tool depends not just on knowing when to use it but knowing what it is actually showing you. Stop for a second and think about how the theater manager would want the graph to look if it were helping her achieve her targets. There are two major changes: either a shift up or down for the demand curve and a shift up or down for the supply curve. Something that makes it harder or more expensive to make movies, for instance, would cause that supply curve to move upward. It costs more to make the same product. The manager's goal, then, would be to increase the public demand for the product so the equilibrium price would not drop (thereby cutting into profits).

This is why the price of a movie ticket can often be less during the day than it is at night, even though people are buying a ticket to see the exact same movie. Fewer people demand to see movies during the day, so in order to sell tickets, a manager of a theater sets the prices lower. At night, many more people want to watch movies, so the prices of tickets can be set higher.

This kind of tug of war between demand and supply curves that are continually shifting up and down as conditions change is going on behind the scenes pretty much everywhere in an economy—

The mathematics of moviemaking

You would probably be surprised at the amount of math going on behind the scenes and under the surface of successful movies.

One of the ways that happens is in getting investors to support the project. To film a movie takes a pile of money before it ever gets to the theater for you to buy a ticket. That means the people who want to make the movie have to rely on investors. These are usually big companies, and in order to decide if this movie is a good investment, they run its variables through computerized finance models to create an estimate for how long a movie will be profitable and by how much. The computers churn through tons of equations, sounding out the different variables: How big of a box office draw is the star? How long is the film expected to take for shooting?

The second part has to do with what researchers have found are certain mathematical patterns that happen in successful films. These patterns have to do with our attention span, the length of some shots in the film, and even the pacing (how often we go up and down the action roller coaster during the film). Good films, it turns out, have a fractal form similar to snowflakes or air turbulence. Cool!

how much it costs to get your car fixed, your teeth cleaned, your late-night pizza addiction. It is even happening as you stand there staring at the approximately 5,307 billion types of breakfast cereal on the grocery store shelves.

An interesting factor that compounds the tug of war process is that sometimes the factor affecting the supply curve has to do with the business owner's choices regarding game strategy against his competitors. For instance, the coffee shop might get a slight discount on some of the coffee it buys, and that would normally shift the curve, and the price, downward. However, the owner is concerned about sparking a price war with other coffee shops in the area, so they might hold off and use the extra profit to do something else that builds customer loyalty.

Grades and Test Scores

Sometimes we are just up against the need to be able to portray a small group of results—such as scores on a test or project. One of the simplest forms of data model to create is something called a line plot. This is done by first determining a scale that would be appropriate and would be able to include all the relevant data points we are working with. Then we would just go through and mark a simple X above the appropriate data value along the x-axis. The result would be a collection of towers of X's of varying heights.

One good example where this sort of graphic model comes in handy is in teaching and grading because the sample size is usually thirty or fewer students in a given class.

Even before we do any number crunching like finding the average, calculating how far off the average most people fell, and so on, think about the general sense of the results that are provided. Would you say this class did well on the material? Or does the class need to increase the time spent on review?

Then we need to consider some choices from the wide variety of other tools and models that are part of a good teacher's

Test Scores

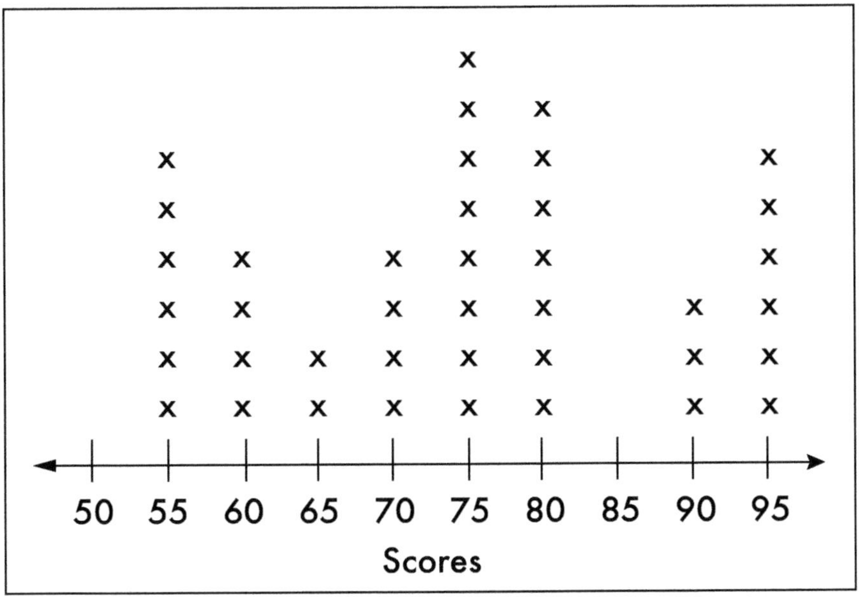

toolbox. These models help track the data and communicate with students and parents. They can help students set achievable goals for improvement.

Just because we used the school example here doesn't mean it has to have anything to do with school activities in order to be useful/interesting. Anytime you are trying to make a change or learn a new skill, this is a great tool to lean on because it is really quick and easy to just plop down on a piece of paper without stressing about your drawing skills or having to spend five minutes waiting for your computer to load up some software. Just a pen and piece of paper—how many hours you spent practicing piano each night of the week, the number of compliments you gave to people today, maybe how many times you managed to sit down and do a few minutes of meditation or yoga, or perhaps instead how many hours of TV you watch every day. This graph can help you make

Meet Steve Hamilton

Please meet Steve Hamilton. Steve is a principal of a K–8 school in Hartford, Wisconsin.

Thanks for talking with us, Steve. Can you tell us how long you've been a principal?

I've been a principal for almost four years now, and before that I was a teacher for twenty years.

Are there some kinds of math used in your role that might surprise people or that are different than you thought?

You know, it's surprising, but I end up doing math that would normally be divided up among different jobs. There is the stuff that goes with running a business, like calculating the payroll, forecasting the budget for next year, and so on. Then there is the stuff that you might think of as part of an architect's job, like a recent renovation of our middle school science lab, determining square footage of rooms for carpet, furniture, etc. There's restaurant management when I sit down with people to plan out hot lunch costs and options, discuss repair/replacement of kitchen equipment and cafeteria tables. And then, of course, there's stuff you think about teachers doing, like analyzing test results, comparing grade performance year to year and class to class. It's funny, some days I even have to do the weatherman's bit by calculating the temperature and wind speed to figure out if the wind chill factor means we need to stay in for recess.

Is there a kind of number/math puzzle you find interesting or that sparks curiosity for you?

Just about all math puzzles interest me. In fact, all through school, math was my favorite subject!

Do you feel like understanding national averages and ***standardized tests*** *helps your students and teachers set realistic goals?*

No, we focus more on the individual student needs using formative **assessments**, as well as longitudinal tracking of each child. We're still doing some of the same math behind the scenes in order to understand what healthy growth and progress might mean from one year to the next, because that allows us to see if we need to provide additional resources to help this or that student reach their potential. Instead of comparing a student's results to a national standardized curve, though, we're comparing them to how they did last year.

That's a cool way of solving the puzzle for meeting each student where they're at. We appreciate you sharing your ideas.

decisions for whatever it is that you want to be a little better/a lot better/the best at.

National Testing

As we move from the small classroom setting to a larger scale, like state and national standardized tests, the line plot is no longer sufficient for the additional math processes that we have to run. In order to provide a tool by which the performance of this student or this school can be compared to the performance of other students and schools, we need a set of data that allows us to find a shortcut to comparing School A against School B, School C, School D, and so on, as well as B versus C, B versus D, etc. We can map the larger set of scores into a computer model and calculate out some key **components**. For example, we might ask: How far below average would a school be before we consider it as needing direct intervention? We are then in a position to easily analyze the numbers and make adjustments to the model. In the end, we are provided with a **substantial** amount of feedback, and that feedback allows for some adjustment in what a fair test of the material would look like.

As you can see on the next page, this shape is called a bell curve. I'll bet you can guess why!

It is easy to imagine to ourselves that this is just weird school stuff that we don't have to think about out in the "real" world. The further along you get in school, the more you might be tempted to think like this. However, this same kind of model is going on behind the scenes of tons of places besides just school tests. The same ideas of the bell curve apply to qualifying exams that doctors must take to practice medicine and the exams accountants have to take to be qualified to help you with your taxes. They also apply to admissions tests that decide who gets into which colleges.

Package Delivery

One of the coolest kinds of problems comes up in situations

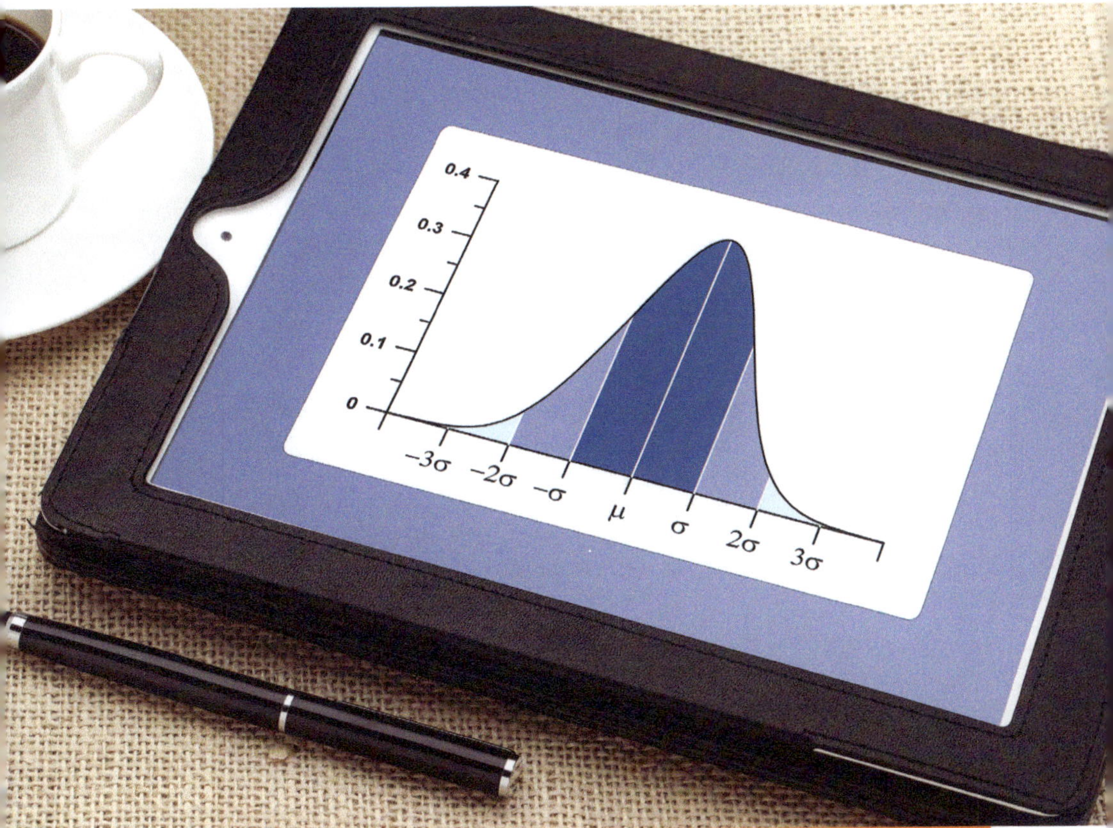

Curves like this are a common tool in statistics. They help us visualize some of the relationships and patterns among a group of numbers.

where we are looking not just for one answer, but for the best among several possible answers. The "traveling salesman" problem is a famous example of something that we might not normally think of as a math problem to start with. The idea is built around this question: Given a certain number of cities and knowing the distances between each of them, what is the shortest possible path that would take me to each of the cities just once and then bring me back to the city I started from?

This problem gets solved every day millions of times over by companies that specialize in delivering packages (or shipping loads of merchandise). They solve it at the scale of moving from city to city, and they also solve it at the smaller scale of moving from address to address within the city.

This same problem-solving approach applies to other fields as well, ranging from production lines in a factory to worldwide travel. It even turns up, in a modified form, in certain puzzles of biology—like DNA sequencing. In these cases, the "city" may be, for example, customers, or sections of the DNA, or it could just be cost, rather than travel time.

The different versions of the puzzle can be **translated** to a simplified kind of map, like the one shown on page 43, which helps clear away distracting or irrelevant detail. As you might expect, the solid circles represent the cities, and the lines connecting them equal some version of cost (maybe time, maybe money, etc.).

Once it has been set up this way, the problem can be tackled using a set of techniques referred to as linear programming, though they are not actual pieces of computer code. Rather, the situation is set up as a series of equations along with some requirements for the possible solutions that the computer comes up with.

Some of the other key models that would come into play in solving and representing the math at work in package delivery would be common business tools, like line graphs to track sales totals, and **demographic** data to help iron out what your customers might be looking for.

Think about this model for a minute, and ask yourself what are some other places that this same kind of thinking applies and helps you come up with the best possible option. One of the first ones that comes to mind for a lot of people is how this applies when you are taking a plane trip (the airline is shipping you, instead of you shipping a box of stuff). Another interesting case occurs anytime you are looking for ways and conditions to save money (trying to get costs as low as you can) or to earn more (how much time you should spend on one job versus another).

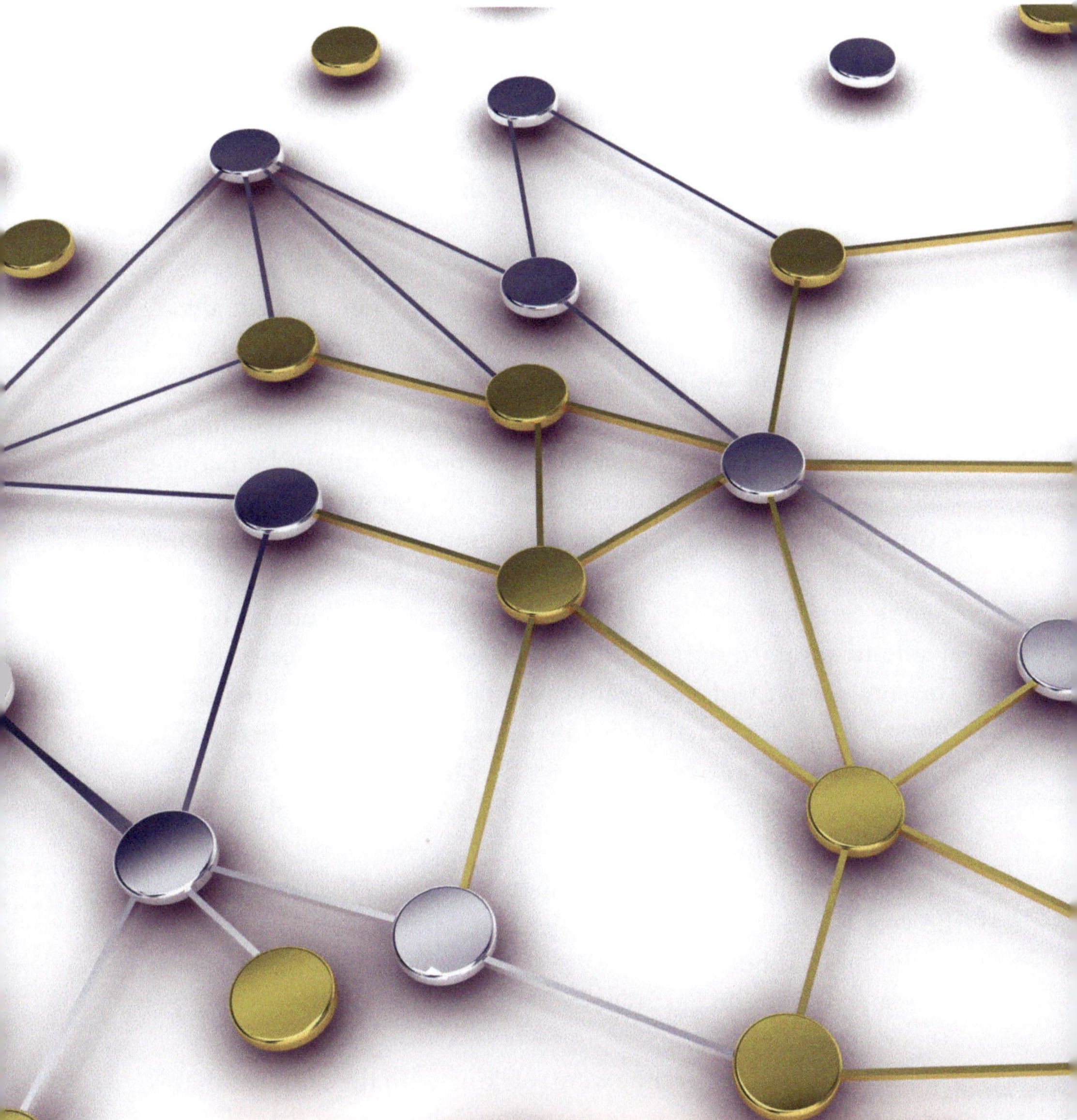

Such simple models of a network can be used to represent everything from the spread of disease to sending information over the Internet.

Political Campaigns

What happens when we are trying to make sense out of something involving variables that are hard to put numbers on? Much of math and the data tools that go along with it work from the perspective of being able to model everyone as if they were unemotional, rational decision makers.

Let's examine a scenario where we're trying to understand what we are seeing happen in a political campaign. It's not always simple to make up our mind about who to vote for based on what they say because we also need to understand why a candidate might be saying (or not saying) what they're saying. Avoiding an issue that you thought the candidate stood for might be because their position has changed or it might be a strategy play to avoid setting up the opposing candidate to jump in and score a big boost in the polls.

We can represent the situation with a matrix where each party is able to support, oppose, or avoid a certain issue (see page 45). The key element of being able to analyze the best outcome, of course, is the work that goes into assigning the estimated payoff mix under different possible results. Here, the numbers in red are the payoff for the Red Party, and the numbers in blue are the payoff for the Blue Party.

In this table, each of the numbers represents the percentage of the vote that each party would get by taking a stance of an issue. For example, if both parties were to come out and actively oppose the issue, Red would get 30 percent of the vote and Blue would get 70 percent of the vote.

Now both parties, if they were working with the same estimates, would be able to see that no matter which strategy Red chooses, Blue will be best off choosing to oppose the issue. That's because the best outcomes for Blue are all in the middle column.

Red Party \ Blue Party	support	oppose	avoid
support	65 35	25 75	75 25
oppose	75 25	30 70	80 20
avoid	65 35	35 65	45 55

Red's best strategy at this point would be to avoid the issue (choosing the bottom row). Even though they are losing voters on this item, they are minimizing the potential damage to their cause.

An important feature of this particular version of payoffs is that Blue's strategy would be the same no matter what Red picks, so they do not really need to carefully strategize on this topic. That also means they don't need to know what the other group is doing before being able to make their choice.

In this case, it is fairly easy to put our fingers on what elements or data would have to shift in order to see the outcomes move to a different place on the matrix. Red would need to change the mix of payoffs in that middle column, either by swaying public opinion on the issue ahead of the voting or by undermining public confidence that Blue would follow through on the issue. Similarly, they could work to push Blue into an avoiding strategy, because Red would get better results from any result in the right column.

The next thing to ask ourselves is what some other useful data tools would be in working to solve our puzzle about what the true position of either party might be on the issue. What do you think

It might not look like it, but these people are boosting their math scores, adding years onto their lives, and making their brains faster.

A new kind of relativity

A half hour does not always equal one half hour.

There have been a number of scientific research studies over the years that show different ways that watching a lot of TV has a negative impact on your brain. One study found that each additional hour of TV watched was tied to decreased classroom engagement, decreased math achievement, and increased chance of being bullied, among other things. A different study showed that kids who watched more than two hours per day were 1.5 to 2 times more likely to have problems concentrating or paying attention.

In contrast, exercise has been shown to improve mood, boost the immune system, and increase cognitive function—including the formation of new tissue (neurogenesis)! Different studies in recent years have also shown that improved aerobic fitness more than doubled students' chances at passing a math test and that it can increase the speed and strength of synapses firing.

So let's think about this: if you opt to not watch TV for that extra half hour today *and* decide to get in some exercise instead, you are earning double bonus miles in becoming happier (and mathier).

would help us see past the fog of campaign strategy? One good tool would be a frequency measure of some kind (scatterplot, bar graph, etc.) to see how many times the candidate or party has voted for or against the issue in the recent past.

Now that you have this idea in your head, think about other ways that this kind of setup could help you make sense out of the best strategy for moving ahead. What about using this kind of scenario to think through options to build up your résumé of achievements before applying to college? You're trying to outsmart many other players. If you all pick the same activities, none of you stand out, so the payoff is lower, right? Should you pick the extracurricular activities in your strongest areas, or should you focus on those where you are the strongest one in the group? What strategy gives you the best advantage over your nearest competitor?

Daily Activities

Coming in as probably the third most familiar way to represent data, right behind line graphs and bar graphs, is the pie chart. This is a circle that has been divided into slices—each of which represents the percentage of a given item or category in a set. While they are common, pie graphs are actually fairly limited in their usefulness beyond a basic sense of **proportion** among items.

It is an interesting feature of our psychology that even when we are not actually looking at a graph or chart, our mind is processing information as if we were. If you think about it, you might actually have noticed times that you were sort of imagining a pie chart in your head when you thought about stuff on your projects list or (especially) when you thought about the clock and how long something would take.

Pie charts allow us to evaluate the relative portions of your time that are spent on various tasks, and then evaluate and make changes if needed. Oftentimes our thinking would be more effective

Pie charts are so useful and so easy to process that it's almost impossible to get through the average business meeting without seeing at least one.

if we would take that unconscious pie chart process and make it conscious. Think how clearly it could stand out to people in the US that in 2015 they were spending an average of about 2.8 hours per day watching television. That comes out to 17.5 percent of all your waking hours, even though some would say these are often mediocre programs. Think how much of a straightforward impact it could have for them to see that they spend 35 percent of all their free time on that one task during the week.

In this way, we see quite vividly how models can take the same information that we might read in a list and make it clearer and more forceful. I won't hold your hand this time. I want you to think about which of the other types of data models we tend to use as we carry on with our day-to-day tasks.

This is a good opportunity to bring something out into the discussion when evaluating a chart, especially when considering what we need to (or should) do in response to a given chart. Let's say your chart shows *x* percent of the week spent on exercise, *y* percent spent doing stuff with your friends, and *z* percent of time spent working on homework. What should you do in response?

Consider the difference, though, if you compared this pie chart with another that mapped out how important different goals are for you, and then started comparing the two charts to see where the biggest matches and the biggest mismatches occurred. Now you have meaningful context and you could say that these graphs show (or don't) a specific problem you need to address. Similarly, we could set up a comparison between current pie chart wedges of what we've actually been doing and wedges from a pie chart that shows how successful people divide up their time instead. That could also give some interesting and meaningful context.

Let's get a little creative here and think about ways we might be able to get more information out of one pie chart, which would make it more useful. The width of the slice tells us about variable *x* (like number of customers per day in a restaurant). What if, instead of choosing random colors, we used shades of gray to tell us about variable *y* (a darker color means an average customer spent more money than lighter colors)? Then, for a third variable, we could make one slice taller than another (in a 3-D version). Maybe this one is for how long the customer stayed at the restaurant before leaving. We could add another variable by having it stick out longer than others, as is done with polar area graphs. What else? What could we do to the pie wedges to add some another variable, like average number of people at a table?

Cars

It's a fact of life that the "stuff" we own—our computer, our car, our refrigerator—breaks. As suggested earlier in the book, it is funny how we think of mathematical equations as models, and we think of computer models as models, but oftentimes we still don't think of things like meters or gauges or electronic machines as also being (or including) a visual model of some system beyond itself.

When we think of cars, most of us are more likely to imagine a car and our relationship to our cars as a small collection of basic line graphs: insurance payments and car payments each month, how far we traveled in *x* amount of time, and so on. Next on the list, even though you might not have realized it was a mathematical model until we just talked about it, the other way we would tend to imagine or frame ideas about our car would be with that sort of vague map and the traveling salesman problem. Should I go by the store first and then the library and home, or would it work better (faster, shorter) to go to the library first?

But when your car breaks, a mechanic will often hook up an oscilloscope to your car to help diagnose what exactly might be acting badly. This machine, shown on page 54, creates an animated sine wave moving across the screen. That moving wave represents the electrical signal passing through some part of your car's engine, and it allows the mechanic to quickly zero in on the problem.

The happiest man in the world

If having a happier life is on your to-do list, Matthieu Ricard is the guy to listen to.

Ricard is commonly referred to as the "happiest man in the world." He has participated in hours and hours of MRI scans as part of a research project. During the research, scientists measured the highest levels of upbeat brain activity. After Ricard put in more than ten thousand hours of meditation (imagine five hours of peace and quiet every day for five and a half years), the MRI scans show his measurements at about 50 percent higher than the level normally considered the high end of the scale. His scans also show negative emotion levels so low they barely register.

Ricard did not start out this joyful and insightful. He actually earned a doctorate in molecular biology and worked for a period before he eventually left his career and moved to the Himalayas to live as a monk.

When asked what his secret is to finding so much happiness, he points to studies that back up his idea that kindness leads to happiness. They have shown that people who give have the highest level of positive results.

Tools like an oscilloscope translate patterns of electrical impulses into a graph that helps us see patterns more clearly.

What other data would we need in order to make the information from the oscilloscope meaningful or useful? By itself, that fascinating little animated sine wave doesn't spell out words like "fix the car." Some line graphs stand out, as suggested, but what about a scatterplot showing when you did different car repairs over the last two years along the x-axis and showing how much each repair cost you along the y-axis. This would allow you to look at whether the repairs have been happening more and more often, but it would also allow you notice if that trend just keeps going up and up or if it tends to go up and then come back down again in a cycle.

I wonder if we could think of other things where plotting them as up and down cycles could help us do a better job of predicting what the values would be at time *x* or *y*. What about the regular pattern of the temperature cycles for us to predict the length and conditions of the growing season? What about adjusting the **cyclical** frequencies of different instruments when we're working with the sound engineer to mix your garage band's first demo CD?

Your Manager Comparing You to Your Coworkers

Maybe you hold down a job. In fact, maybe you hold down a job at the same movie theater where you lined up for tickets and candy earlier. (What's the fun of a night off if you can't go hassle your coworkers by pointing out that you are off and they aren't?) It might seem like simple bar graphs would work to compare you and your coworkers on different aspects of your job performance. One bar graph could show how much time it takes each employee to keep his or her station clean, and another could show how long it takes you to stock supplies at the end of your shift.

We discussed stacked bar graphs earlier, but there is an additional kind that is mapped to a 0–100 scale. If you want to compare values of some variable from one month to the next, or from one group to the others, grouped bar charts can help make those comparisons more clear. They can make it less clear, however, if you want to still be able to see the totals in one month compared to another. Switching to a stacked bar chart reverses this. It can make it easier to see what the total is, from all variables, from month to month. In this case, though, it is less clear how each part of the stack compares to that part in the other months or

groups. One way to solve this is to convert the items on the stacked columns into percentages.

By setting the time it takes different employees to complete tasks to a common percentage scale, the manager can get a much more accurate comparison of how well employees are using their time. (Does one person spend a disproportionate amount of time cleaning?) If one cashier works 6 hours and one works 7.5, then a regular stacked bar graph would make it harder to fairly compare the two employees, but a stacked 100 chart would work pretty well because the hours have been converted to percentages. On such a chart, 30 percent of 4 hours would be the same size as 30 percent of 6 hours. Naturally, this only works for job tasks that take more or less time in proportion to the length of the shift. For example, this might include counting the cash drawer. It would not work for something that takes the same amount of time regardless of shift length, such as opening the theater doors.

This analysis isn't limited to just your manager, though. Maybe you become team lead at your job, or crew chief. At that point, *you* are the one doing the manager's role of applying this kind of performance comparison to evaluate employees and to make decisions about the schedule for next week. What are other places you might benefit from applying this same kind of metric? What about doing comparison shopping between computer brands? You could use this approach to see how much time or money is lost because of the percent of time Brand P is down from crashed hard drives and then compare that to Brand M.

Meditation and Brainwave Shift

Here is one of my favorite examples of being able to turn our technology inward and use it to understand some things about ourselves. In this case, we are talking about the very brains that built the technology.

By hooking up tons of **electrodes** all over our brains, scientists and researchers can carefully monitor the electrical signals being generated by our brain. One of the most interesting things they found from doing this is that veteran **Buddhist** monks can quickly enter a meditative state that changes the frequency of those brain waves. This state also lowers their breathing, their heart rate, and the presence of stress chemicals in their system. Research has shown that meditation can also give your immune system a big boost, strengthen your memory, and help beat depression—often better than prescription antidepressants. Good news: you don't have to go off to a monastery (unless that's your dream life). Scientific studies show that the brain and body benefits start to kick into gear after even a couple weeks of short daily meditation.

What are some other ways that monitoring and modeling our brain waves, heart rates, or our attention spans could be used? What about connecting back to our opening topic of fitness and nutrition? We know that muscle building and aerobic capacity follow up and down cycles when we are training, so being able to

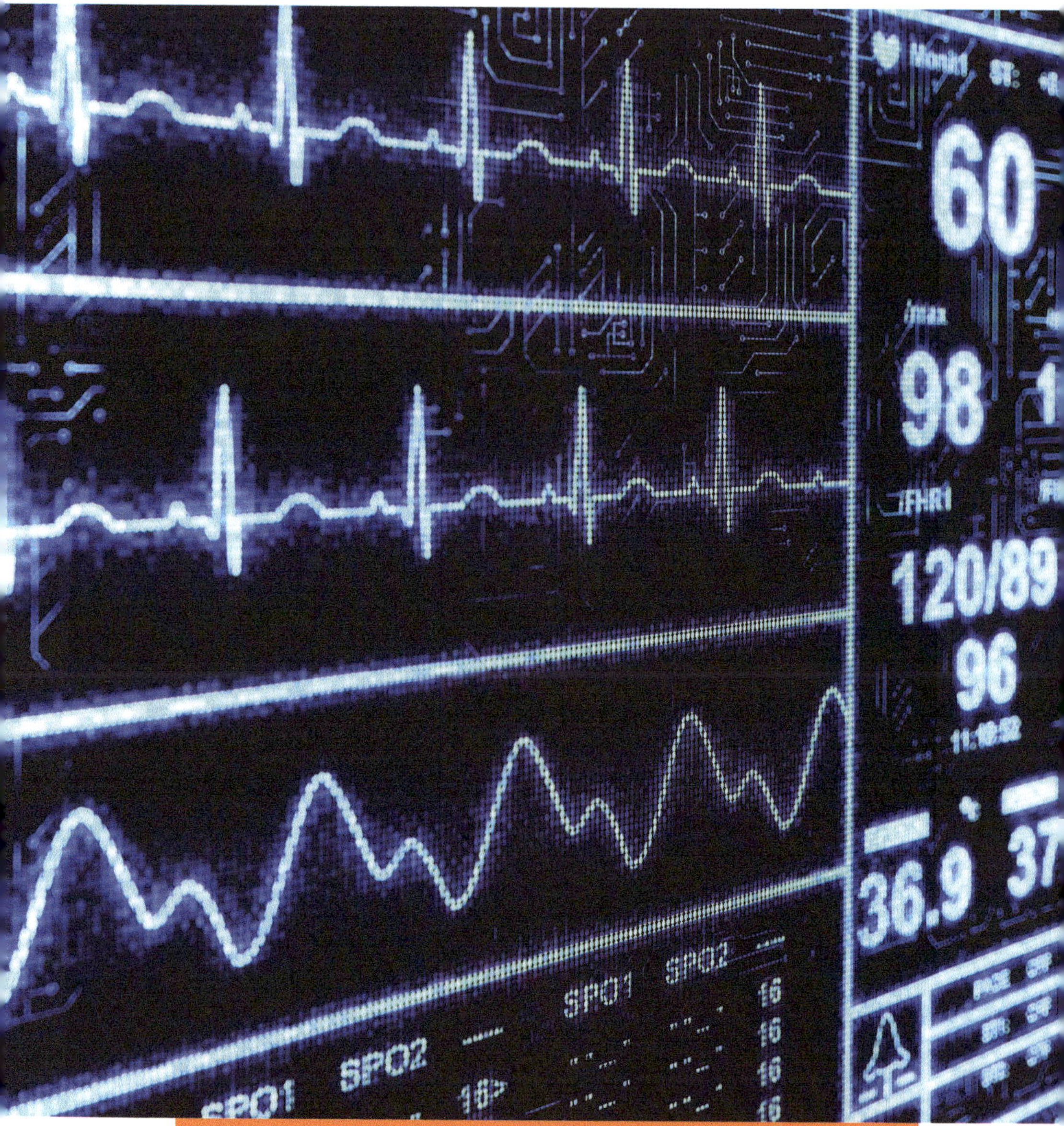

Different parts of our body and brain generate different patterns of electricity that can be modeled by medical equipment to look for irregularities.

map and predict those highs and lows could help us plan when our training should be more or less intense. Similarly, our ability to study and learn new material follows up and down energy cycles, so mapping and using that kind of information about yourself could be a good way to improve how you use your study time.

Gym Memberships

As our next example, let's consider a situation where you are thinking of signing up for a gym membership. During your first visit to the gym, you will probably be forming a kind of vague idea in your head of different types of people you see there, their age categories, and so on. It's part of the process we go through, often unconsciously, to gauge whether we fit into a given group or environment. I say "vague" because you will probably not actually be imagining little vivid, colored bars on a page in your head but will instead be assigning general tags like "loads of bodybuilders" or "barely a handful of women."

Interestingly enough, in order to figure out how to build his business, the gym owner goes through a similar process, except he then takes the next step and starts putting those observations into a graph where he can see them more clearly. He will update that graph as memberships add or drop, and he can use it to help with decisions about where to concentrate his energies for marketing to help grow his client base.

A histogram is often confused with a bar chart (or column chart). There are two key differences, though. The first is that while a bar chart can have categories or kinds of things (lions, tigers, bears), a histogram is dividing up one continuous range of values (ages, heights, lengths of time). The second is that a histogram uses ranges for each category rather than a single, distinct value.

So, when we go to build a histogram, the first step is to chop the range of values into separate small bins—that is, we divide the whole x-axis into small **intervals**. These are usually intervals of equal size. Then we would proceed to figure out how many items fall into each interval. The intervals are usually set up in consecutive order for the sake of clarity (such as youngest to oldest). The intervals should touch at the sides. Right where one leaves off, the next one picks up. Each data point is within one of the intervals.

As an example of where we would want to look to a histogram to organize and clarify our data, consider the case of a health club that wants to see what age groups they are and are not successfully appealing to with their advertising campaigns.

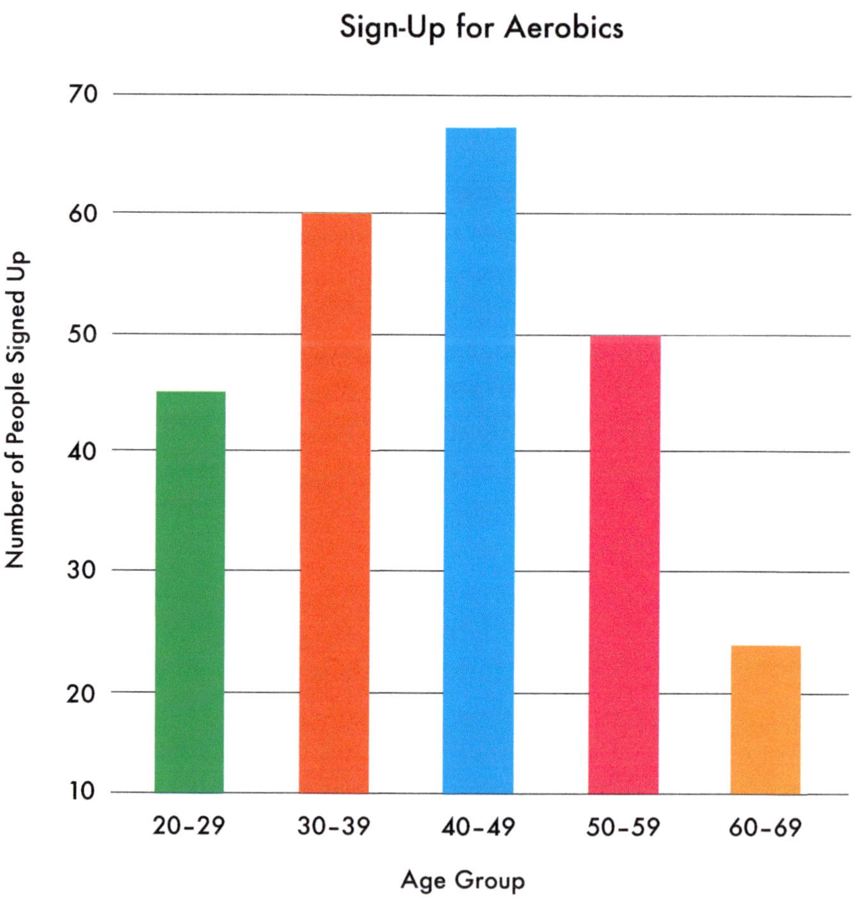

Compare this with some of the other cases we've looked at. What other data models might the gym use to consider how it is doing and where it should focus its efforts? What are metrics that potential clients, like you, would be doing in their heads as they think about whether it would be a good gym to join? Is having a lot of clients in the younger age range better, or is it preferable to have a lot of clients in the older age range? To solve questions like that, we might want to look at things like: In which age range do members tend to do the most networking to convince friends to sign up? In which age range are members most likely to sign up for additional services like personalized coaching, nutrition supplements, and so on?

Now I realize this looks a lot like a nutrition and exercise type of question, which we've hit on in earlier sections, but it really isn't. This question is about fitting into a given group. You might well use the same model if you were thinking of joining a community choir. Maybe you are looking for a good meditation group to join so you can get all those awesome peace and joy things working in your favor. Different groups with different demographics can be a better or worse fit for you, so you could use this same approach.

Martial Arts Instructors

Some situations that would seem like a natural fit for a line graph run into a weird kind of problem. Take the example of a tae kwon do school and its instructors. Let's say they like to have a ratio of 1 assistant instructor for every 10 people in class to help out with techniques, forms, and holding pads for striking. In a normal line graph, that would suggest that if there are 13 people expected, they should have 1.3 assistant instructors on hand. But it's very hard these days to find 0.3 of an instructor!

Instead, what is called for here is something known as a step graph. This is a special version of a line graph. Instead of one straight line, a step graph is made up of a series of short, disconnected, horizontal lines where the left edge of each one begins exactly above the right end of the previous step.

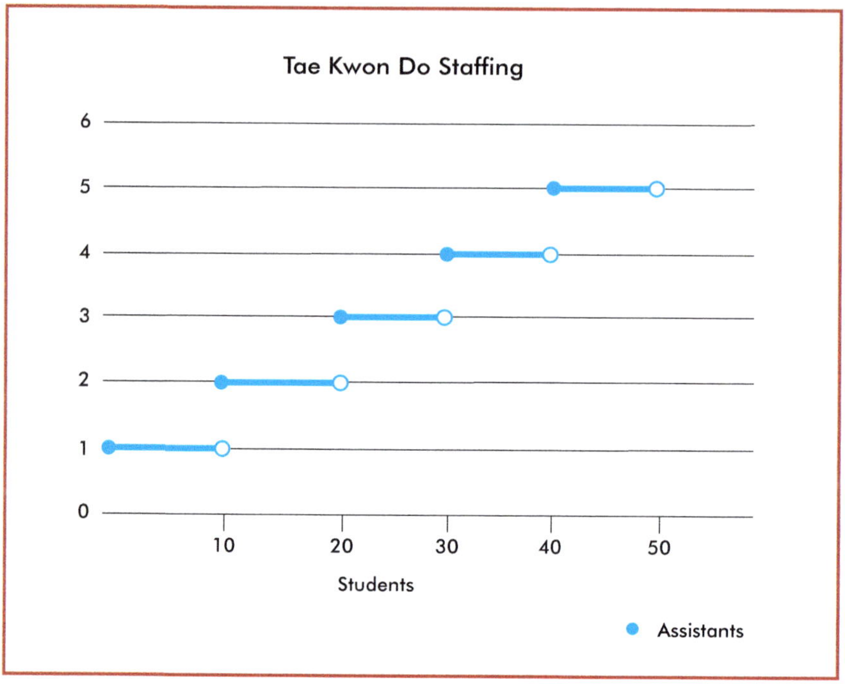

As a reminder, the open circle means that point is not included on a given step. A closed circle, however, does include that point on the current line.

So, this tells how many assistant instructors to schedule given a certain number of students. The most likely partner graph to this would probably be an attendance forecast of some kind—maybe a bar chart showing average attendance on different days of the week for that scheduled class time. How would you adjust this graph to show assistant instructors needed for kids' classes if the goal there was one assistant for each five students instead of each ten like the adults class?

Understanding how this chart works comes in handy for jobs where the staffing depends on some volume of something (customers attending, cars to be repaired, you name it). It also applies to just about every kind of event or volunteer activity you would ever be involved in—from cleaning up along the river to working at the soup kitchen to organizing a fundraising banquet. Beyond that, though, it also applies to supplies or equipment you would need to plan for. How many buses should a teacher order for the field trip (since he can't have one-quarter of a bus)? How many cups and straws to order for the movie theater concession stand—you can't order one-third of a straw, right? How many new tires or spark plug wires or brake shoes do you need for your car repair?

Hospital Illnesses

As suggested in an earlier example, there are some forms of data that are almost right for a pie chart, but some key information would be lost or distorted if we tried to fit it into a conventional diagram. Instead, there is something called a polar area diagram. While the angles of the sectors are divided out as a percentage of the total data set, the radius of each slice of the pie can be different. For example, if 10 percent of the patients in a given month have viral infections, then the colored segment for that would be 10 percent of 360 degrees (36 degrees). Then, if the average stay for those patients is, say, ten days, that slice of pie would stick out to a radius of 10 (centimeters, inches, or whatever unit of scale is being used for a given chart).

Let us imagine that this is being used to model illnesses and care durations at a given hospital. Each sector would have an angle measure reflecting the number of cases of each illness in relation to total illnesses in the hospital. However, not all illnesses need the same amount and duration of care, so we can let the radius equal one of those (total costs or average length of hospital stay).

By way of an interesting side note, this is part of the logic in the

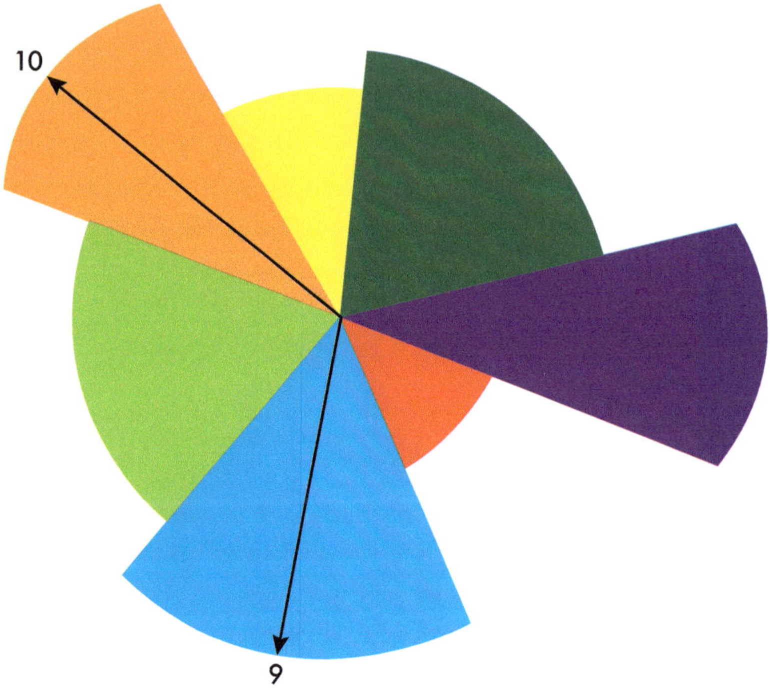

Sample Values Shown:

- Zombie Plague: 10% of patients. Average stay 6 days.
- Viral Infections: 12.5% of patients. Average stay 10 days.
- Cafeteria Lunch: 15% of patients. Average stay 7 days.
- Alien Spores: 18% of patients. Average stay 9 days.

chart that was put into widespread use by Florence Nightingale.

Aside from describing the number of cases by type of illness and length of stay, what other key metrics and tools could be working behind the scenes when you visit the hospital for that concussion you got in tae kwon do class (because you just don't keep your left up no matter how many times Master Lee tells you to)? What are some things that would affect decisions about providing quality health care? Do some procedures cost the hospital more to perform than others?

At the same time, let's consider other types of applications for this polar area chart. The circle graph layout makes the most sense when we are comparing similar things—models of cars, types of illness, attendance at different professional sports, etc. The extension along different-length radii makes the most sense when there is another variable that has significant impact on comparing the different groups. Jewelry purchases might be another interesting example, because there are different numbers of precious gems sold but also huge variations in prices of different gems.

Flying, Paddling, Sailing, Sledding!

The last kind of model I want to look at in this chapter has to do with the way different forces are usually acting on a given body. It could be the wind versus plane engines, or a current versus boats. These forces have strength and impact the body from a specific direction. They can be modeled by vectors.

Each different vector is represented by an arrow. The length of the arrow is the **magnitude** (strength) of the force pulling on the object. The arrow points in the direction the force is acting. An important rule for working with these is that two vectors are the same if they have the same length (magnitude) and are pointing in the same direction. This means that if we move an arrow (a vector) to a new place on the page—as long as we haven't rotated it or flipped it—it is the same vector we began with.

To give a quick example, imagine that you are trying to paddle across a river in your canoe. The force and direction of your paddling would be straight across, so if there were no current, it would make a nice straight line. On the other hand, if you went out to the middle part of the river, with a current, and stopped paddling, the force and direction of the river would be an arrow pointing you straight downstream (north on this map).

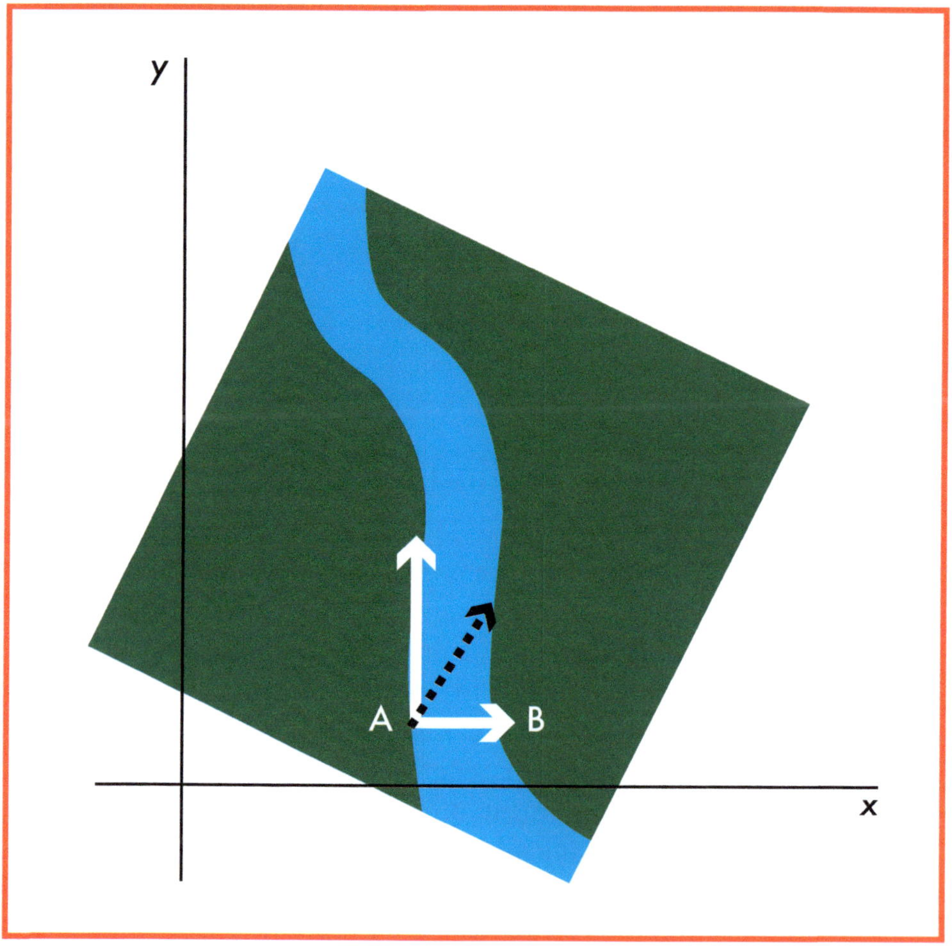

Even if you don't have time to do the calculations in the moment, understanding the math models of your sport helps you make split-second decisions.

When these two different forces are both acting on the same object (your canoe), you will move at an angle somewhere between 0 degrees (going straight across) and 90 degrees (going straight downstream). How close to 0 or to 90 depends on how hard you are paddling and how strong or weak the current is. It should make intuitive sense to you that you would likely land on the opposite shore somewhat down from the initial target—shown by the dotted black arrow.

In the example illustration here, the current is quite strong—indicated by the long arrow, instead of a short one—so you will end up a good ways down from your desired target. Of course, most of us would switch the direction of our paddling to compensate, but the type of effect would be the same. It would just be a contest between the direction and force of the river's current and the (new) direction and force of your paddling.

We would reach for this kind of model when we are discussing things like flying or sailing against a strong wind, but it also applies to situations like hurtling down the hill on a new sled where forces are pulling us in different directions at the same time.

This model need not be limited to physical forces. Take a minute to think about what the equivalent case would look like if we were to switch the frame of reference to talk about something like different business opportunities for our company, or about competing demands on our time (sports versus homework, job versus friends, etc.). We still need to appreciate that the assignment of magnitude is critical. If we only make the force of academic work less than the force of hanging out with friends, we might make some critical mistakes in direction and find ourselves fairly far off course.

Conclusion

That should have felt like a bit of an adventure. There are many meaningful charts you use in working to succeed at any of these areas of your life that we've talked about. If you update them often and make decisions based on them, the better your success will be. Just consider the point of using charts and models to start with: they help you see the patterns and trends more clearly. They help you include important pieces of information in a way that's easy to remember, and those things help you make better decisions.

The other thing I want to leave you with is the possibility that charts and models have an additional impact besides just helping you see the data or patterns more clearly. They also fire up our motivation. We are hard-wired deep in our brains to want to get better at what we do. Any time we put down a picture or a map or a scorecard showing how far we've gotten, looking at that and thinking about that ignites our desire to get farther, to jump higher, to run even faster than we did last time.

Do you want to be better at your job and get the raise or promotion? Don't sit around waiting for your manager to show you a measurement of how you're doing—measure yourself.

Do you want to be better at sports or save up for a nicer car or get into a better college? If you are trying to get somewhere and you don't have a map, you're just hoping you will accidentally stumble in the right direction. Maybe you think the universe thinks you're special and will bend around so you end up where you were supposed to. That's crazy talk. Make a map. If the universe happens to offer a shortcut along the way, that's fine. Don't expect the shortcut to bypass the whole journey.

In the next chapter, we'll see how different businesses and careers use some of these maps and models to help themselves navigate toward success. In the process, you will learn about some additional kinds of models that you can use to help you along your way.

Before, after, and during military engagements, a lot of analysis of mathematical models goes on to help the military succeed as quickly and effectively as possible.

THREE

Modeling in Others' Everyday Lives

Let's look at examples from a wide variety of careers to see how different trades and professions use some of the ways we can model information about something going on in the real world.

I'll mention some of the ways they use certain common tools, like line graphs or histograms. But the main focus will be on attempting to show some of the distinctive ways they would use various maps, graphs, and other models. Keep in mind the goals mentioned in the introductory chapter. We're not just talking about models for the sake of models. Rather, I want to show you how professions use models because they make it easier or clearer to see how certain variables might be changing, how there might be patterns that stand out, and how those variables are interacting with other variables.

By taking a tour of the different kinds of models that stand out in different careers, hopefully you will start to get a feel for why those models are particularly valuable in that job. In other words, it should start to make sense to you which kinds of models seem to fit which kinds of problems or answer which kinds of questions.

In addition to looking at central examples and some related examples for each career type, we will also talk a little about how the new kind of model being discussed might play a role in other, similar types of work. In the process, you get to imagine for a little bit taking on each of these different jobs, and you can think about what it is like to spend your time tackling the kinds of ideas and puzzles that go with them.

Military Analyst

A person in the role of a military or defense analyst will be working to constantly update the research on both domestic and foreign military actions, developments, and capabilities. He or she will then collaborate with others to generate meaningful dialogue about the best strategic decisions going forward. To do this, analysts study and produce various maps, charts, and models to assist those key decision makers in forming a best-case scenario.

Hawks and doves

This particular example has to do with nations deciding whether or not to become aggressive and **escalate** to open warfare when two countries are playing tug of war over some particular resource, such as the oil fields in Iraq or the diamond mines in Africa. Because the two primary choices are to be overtly war-like and hostile (like a hawk) or to be peaceable and yielding (like a dove), the game is usually called Hawk-Dove in game theory discussions. The scenario is modeled on page 73. In the model, *R* is the value of the contested resource. *C* is the total cost of having a war.

From our model, we see that if Red Nation picks the strategy of gearing up for war and Blue Nation does the same, then the two sides will end up going to war. There are a couple key **assumptions** built into the numbers shown here. The first is making the split

	Hawk	Dove
Hawk	$(R–C)(1/2)$ $(R–C)(1/2)$	R 0
Dove	0 R	$R(1/2)$ $R(1/2)$

50–50 at the end of the conflict. If the two countries were not pretty evenly matched, then it is likely that the weaker of the two would have opted for the Dove strategy. As a result, the model allows that the two sides would end up with about the same cost incurred from waging the war (half the total cost) and control of about half of the resources that were being fought over in the first place.

On the other hand, if both sides opt for the peaceful settlement, and neither attempts to go for broke by seeking control of all the resources, each side would come out with about half the resources and incur none of the costs of war. Notice that in only one case of the four does anyone incur the costs of there being an outright war. In all three other scenarios, open war is averted.

The military analysts run this simulated game over and over while adjusting the variables each time to see how the situation tilts. What if the war drags on so that C ends up being bigger than the R they were fighting over in the first place? What if one side is likely to get slightly more than half the resources at the end because of some geographical advantage? The trick, of course, is to not only figure out your best (and therefore most likely) payoff strategy, but to figure out that of your opponent as well.

Meet Kevin Freese

Let me introduce Lieutenant Commander Kevin Freese, a US Navy officer with twenty-one years of service (thirteen in the Navy, eight in the Marines before that).

Thank you for joining us, sir. Can you tell us a little about the role math plays in your work?

Well, that's a great question. A lot of what I do in my position is closer to the game theory examples you mentioned, which is a very mathematical process, but people don't always realize that since it doesn't "look" like a pile of equations and symbols.

One of the models is called "four ways of seeing." We ask four basic questions: 1) How do we see ourselves? 2) How do we see them? 3) How do they see themselves? 4) How do they see us? It is a great tool to think through costs and benefits from both (or multiple) sides of the issue and gives you a better understanding of a problem. It can also be refined by expanding it to include other groups, individuals, etc.

Is there an area of math in your work that you find particularly cool or that sparks your curiosity?

What made me interested in math was, while attending Command and General Staff College, I took an elective on Space Operations. Two things happened in the class. First, I learned that you have to set up some problems differently than you would on Earth, because the satellites are moving so fast (20,000 miles per hour and higher), and things moving faster experience time more slowly. Second, and more significant, part of the course was on orbital mechanics. I was really intimidated at first, because I had taken trigonometry but not calculus in college, and calculus seemed like this nebulous, super-scary field that only geniuses understood. However, during that course, I realized it was really just taking the math I already knew and applying it to measuring changes (in this case, changes in velocity). Suddenly it wasn't intimidating anymore.

So is it fair to say that is an area you wish you knew more about?

Every time I learn a little more math, it opens up new questions I want to explore. I've started to think "enough math" is as elusive as having "enough money" or "enough time."

I guess I think that too, but I hadn't put that into words. Excellent point. Thank you again for joining us, sir.

What do you think the outcome would be? What is Blue's best strategy, not knowing what Red will pick? Does it look the same for Red? Are there important variables this model does not account for?

Geographical models

Some of the other kinds of modeling that would characterize the role of a military analyst would include a strong sample of geographical models representing troop placements, supply routes, key weapons facilities, and major military installations. **Supplemental** information like types and numbers of missiles, war machines, and troop strength would very likely be condensed in a variety of bar charts, while information like movement rates and combat effectiveness of a given country's troops would be modeled as straightforward linear equations and sets of equations.

Let's bounce a couple of things around in our heads, then. The first is to ask ourselves what peculiar types of problems a game theory approach helps us see or solve better than other tools. The answer is any time where the issue has to do with people rather than pennies and particles. Game theory applies to situations like the social sciences of politics or economics where the outcomes are not just determined by how fast this train was going when it left New York or how much money that company made in November. It applies to situations like, "What is our best strategy to match the strategy our opponent picked for winning some competitive struggle?"

In that case, what would be some other specific types of careers where this model might be especially useful? Well, we mentioned economics, but let's pin that down to some concrete examples that you can imagine. One might be deciding whether or not to expand a company to start doing business in a new country. Whether or not that would be profitable for us depends on how the businesses in that country react to the new competition and whether other companies from our own country decide to copy us and expand into that country as well.

Another example might be the job of biologists trying to understand how animal populations in an ecosystem will adjust over time to certain changes in their environment. How well each animal species does using its strategy will partially depend on which strategies or adjustments the other animals make. The animals are not consciously trying to outguess each other in the same way we talk about in human game theory, but the outcomes behave "as if" that were happening once the strategies start in motion.

Theoretical Physicist

Theoretical physicists are the people that most of us probably think of when we hear the term "physicist." They include some of the most famous, like Einstein, Heisenberg, and Michio Kaku. Theoretical physicists work to formulate and test theories and **hypotheses** about the nature of the universe. Their areas of research vary from trying to understand the way time and space themselves work (up to and including possibilities of faster-than-light travel), as well as the nature and effect of black holes and dark matter.

Not only do theoretical physicists work to build theories and models, but they also spend a great deal of time sharing and engaging with the research by others in their field. This is true no matter which area you study, from our little solar system to the soap-bubble structure of galaxies stretching around empty areas of space.

Strings

For those of you who have heard of Michio Kaku or Brian Green, you will likely also have heard of something called string theory. Let me give you the basic version. In traditional physics, we tend to think of everything being made of up little, tiny pool balls, and these are zipping around, bumping into each other and bouncing

Here we see an artist's version of how the universe—at the smallest level—is made up of tiny little vibrating strings.

off or forming bonds. However, physicists think that even smaller than these are **quarks**—smaller, super-tiny pool balls—and that those are made up of something even super-tinier. Those super-tinier things are like vibrating strings. They are kind of like wadded up rubber bands, but they are vibrating at different frequencies, behaving more like a violin string or a guitar string.

Hang on to the safety bar, because the roller coaster is at the top of the hill and we're about to drop. There is extra space folded up inside the rubber band itself. The mathematical models that allow us to figure out how gravity and black holes and stuff work tell us that these little bands aren't just lines, like a super, super thin guitar string. Instead, the math tells us that these strings have extra dimensions of space *folded up inside them.*

We normally think about everything in three dimensions, right? Well, these little strings have ten dimensions curled up and stored inside. Take a look at pages 78 and 79 for an approximation of what a cross section of one of those little vibrating strings might look like.

See how there are extra loops and twists on the inside? But you would only see them if you were looking at this unbelievably tiny scale. Think about how your hair looks like a single, simple strand until you look at it under the microscope. It's the same idea, except way curlier than you could even imagine your hair being!

Secondary tools and models

Theoretical physics is, by nature, one of the most thoroughly mathematical fields you could imagine. It is probably second only to something like a math professor in its depth of exploration of the ideas and equations. That being said, theoretical physics has a strong reliance on the ability to construct visual models as a frame of reference, from modeling the bends and folds of space along three-dimensional coordinates to ideas like soap-bubble geometry to represent the clustering of galaxies around regions of empty, expanding space.

Other kinds of graphs and models that would be helpful in condensing and organizing complex mathematical results in theoretical physics include vector diagrams to show the amount and direction of different forces. Physicists are also going to rely very heavily on graphs of functions in two and three dimensions to represent the relationship between different variables and how those behave over time. Perhaps the third most likely kind of model would be portraying things using complex geometric shapes to try to capture the way space and gravity behave.

It might seem like the images and models for theoretical physics are somewhat unique to it and would not really have much use in other areas or other careers, but there still are some. Those careers and areas also tend to be ones that are highly mathematical as well. A particularly cool example, though, is the way that models for graphing and predicting the vibration of rockets turned out to be incredibly useful in graphing and predicting the bouncing movements of stock prices in the investment industry.

The one other area where we are likely to see models borrowed from theoretical physics is in the field of architecture. Some of the interesting ways that space seems to fold up on itself and the ways it changes when subjected to increasing gravity provide some cool solutions to problems in architecture, like how to change the shape of a building to get more strength with less material or to increase its stability when reacting to outside forces.

Stock Trader

Let's come back up to the scale of the world where we can see stuff happening. And, most people will tell you, nowhere is there more stuff happening in a given slice of time than on Wall Street.

The job of stock trader is a high-speed, high-intensity life as they constantly scramble to stay informed on current movements of the prices of stocks so they can work quickly to make the best sales and purchases on behalf of their clients. Even small shifts in the price of a stock from moment to moment can mean a difference of thousands of dollars for the client.

Some of the numerous facets of this career include researching and analyzing the data in the charts and graphs to get a picture of how a stock is behaving, advising clients on purchases and sales of this or that stock, and ongoing communication with the clients to keep them well informed. At the center, though, when all the analyzing is caught up, there is the dangling carrot: buy low, sell high. That is how you make a profit in any business, of course, although in most businesses, the price goes up either because you have made something—like turning the flour and sugar into cookies or turning the metal and plastic into a car—or because you have done something—like stitching up someone's scalp because they misjudged just how good they are at parkour.

With the stock market, some of the reasons the price is bouncing up and down all day are because of something that happened. Maybe a computer company introduced a new program, or a cake-mix company just hired a new CEO who uses a machete to cut through red tape. Some of the reasons, though, that the price bounces up and down are because someone bought or sold a bunch of shares and everyone else thinks there must be some secret reason why. So they also do it, hoping to get the benefit without knowing the big secret.

Open and close

Whatever is happening on a given day, the stock trader has to be watching very closely so she doesn't miss it when the prices start to shift (either up or down). Since stocks that a lot of people are interested in go up and down several times each day, the brokers and traders are always measuring it "compared to where it opened." That just means: What price was it listed at when the stock market started trading this morning?

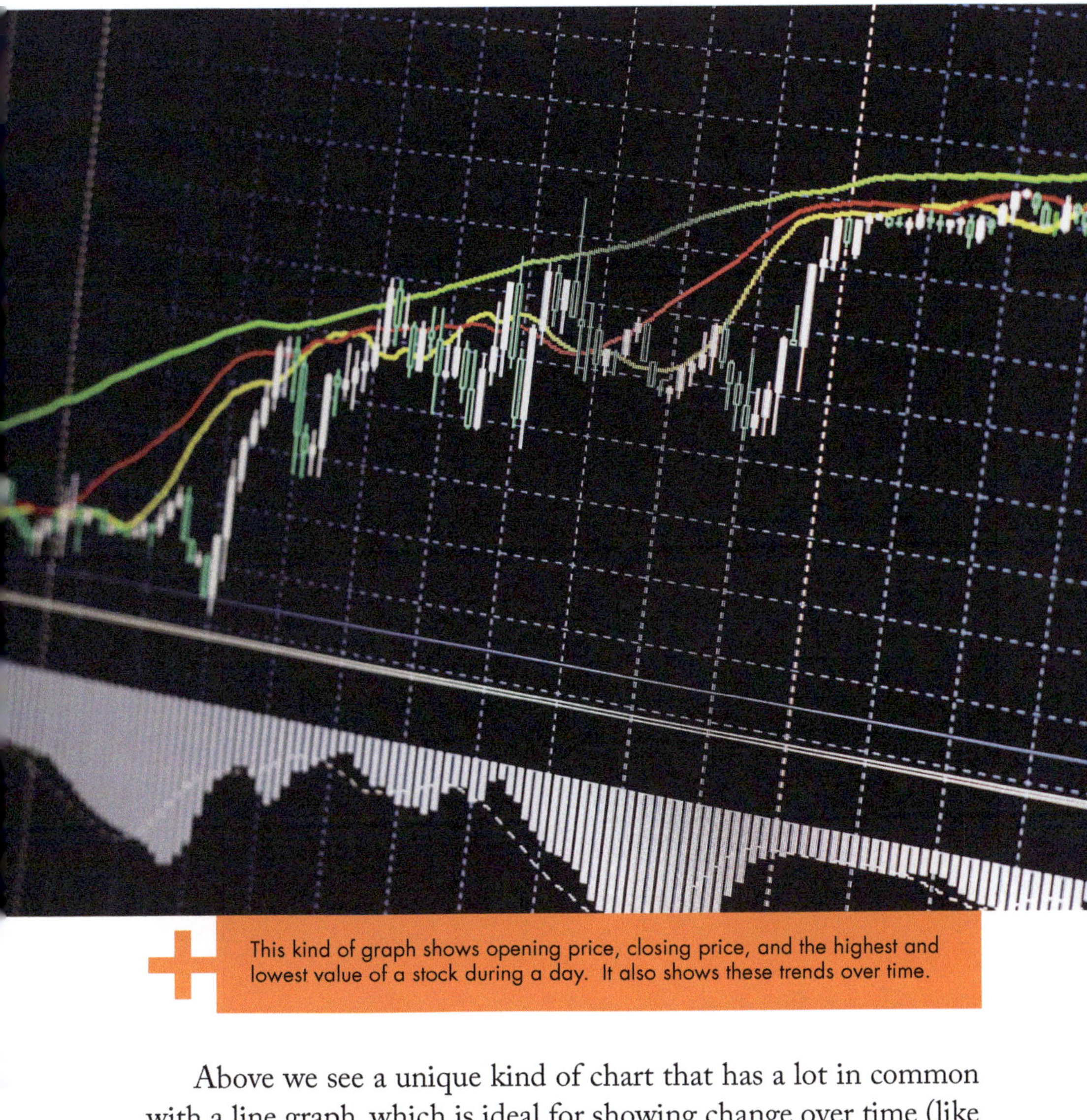

This kind of graph shows opening price, closing price, and the highest and lowest value of a stock during a day. It also shows these trends over time.

Above we see a unique kind of chart that has a lot in common with a line graph, which is ideal for showing change over time (like day to day or month to month). However, notice that at each point has a bar, like you might expect on a bar graph instead. The top and bottom of each bar shows the price where the stock started that morning and what the price was when the market closed. White versus blue here allows us to figure out when the price went up from the morning to close higher and when it went down for the day.

A lot of people see mathematics as a kind of crystal ball, and they are convinced that if we just look closely enough, we can see the future.

Searching for that crystal ball

I mentioned that theoretical physics probably comes in at number two on the all-time list of mathiest math fields. Well, the world of finance is probably third on that list. You're talking about a field where people won prizes and made billions of dollars by being able to apply equations from rocket science to figure out how to predict stock prices. Aside from living and dying by the open-high-low-close charts, like the one on page 83, the trader's next best friend is going to be linear forecast models, and that means lots of **regression** functions to find the right equation(s).

It would be misleading to imagine that all of successful stock trading depends on careful analysis of the opening and closing prices over time. Much of what goes on in the industry is some of the things we've talked about in other sections and with other models. Is the line graph for the profits of this business moving up over the last few months for some business reason or just because a bunch of people have been buying shares? Similarly, a lot of the process involves trying to get a sense of how the game theory payoffs will look for Company X depending on which strategies or reactions Company Y chooses.

Amazing statistical and probability models help traders figure out the forecast for profits, how much they should be willing to bet that they are right, and other ways to invest money to help reduce how much money they lose if their estimate was wrong. Pie charts jump in to play a supporting role in helping stock brokers and their clients keep track of the different amounts of their money invested in a given company or industry. Keeping that balanced is a huge part of the process of putting a safety net under any particular stock they buy.

Sociologist

Sociologist might be a career you have never heard of. A sociologist's job is to increase our understanding of how social groups and societies work and provide us with tools to help improve our society. To do this, their work involves setting up and running research projects and conducting personal interviews to gain a better understanding of the roles and relationships in a given society. For instance: Do societies that treat women as equals grow faster or slower than societies that think men should be paid higher and have more rights? Another quick example might be:

Do societies tend to be more peaceful or more warlike when their political leaders are also outspoken religious leaders?

At the heart of these different tasks is the process of collecting and compiling data, and then analyzing it to look for patterns. Once they see new patterns or gain new insight into how something works, they have the responsibility of putting together the different puzzle pieces into formats to help others easily understand the information and use it in making decisions.

Demographics

One of the kinds of data that is sort of a backbone for some of the other key measurements in a society is the shape of a population. How many men are there in each age group compared to women, for example? How many people of working age are there compared to people at retirement age? Are there enough workers to support the retired people? Similarly, how are those numbers changing over time?

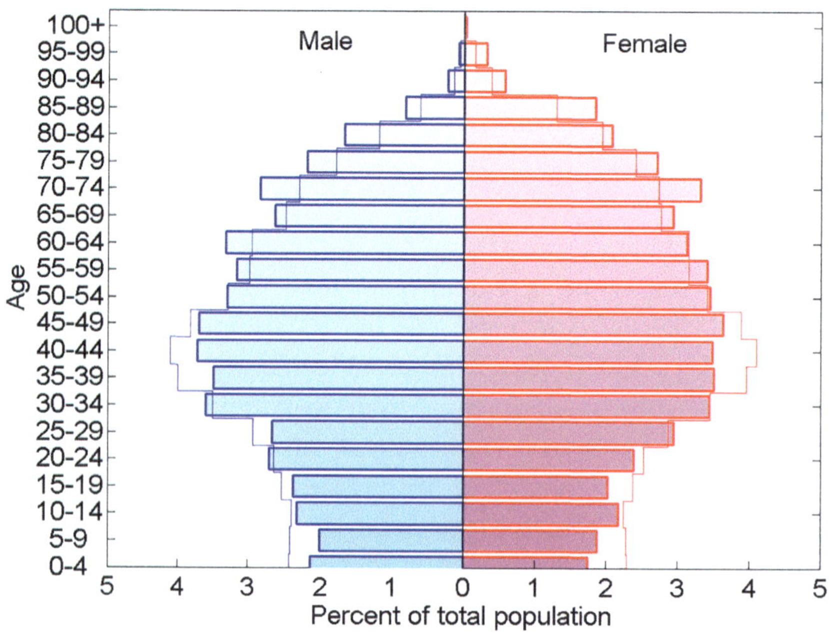

On page 86 we see a kind of graph that is commonly used to help us see some key pieces of information along those lines that would not stand out as well for most of us if we were just looking at a long list of numbers.

Three variables

Another really cool kind of model to represent data is something called a bubble chart. Here we are able to use *x* and *y* in a way we are familiar with, but then, to help us compare apples to apples, this chart also shows the relative population size of the country and it uses color to show which part of the world that country is in.

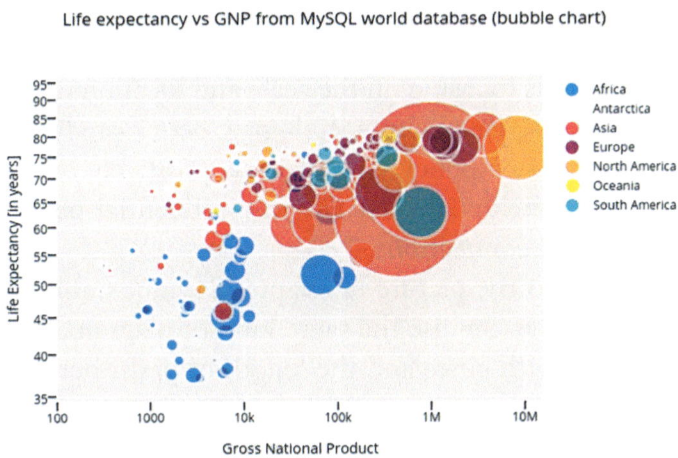

So, for instance, on a normal graph, we could plot points showing gross national product for a given country (which is a measurement of all the stuff the businesses in a country make and sell all over the world each year) along the x-axis, and the average life expectancy of a country (that means the average age at which people in that country die) going up the y-axis. However, with a bubble chart we don't just have dots. We can change the size of those dots, and they become bubbles. The size of the bubble is determined by the size of some other variable. In this case it represents the population. We see the largest red bubble is China, while the orange bubble over near the edge is the United States—significantly smaller in population.

This graph compares GNP, life expectancy, and population size. It generally shows that countries with higher GNPs and larger populations have better life expectancies. We might expect results like this because generally, as a country's productivity and wealth go up, we would naturally expect the country to have better health care as well.

Influencers

All the research on attitudes and behaviors in the context of social and political forces swirling around in a society helps sociologists understand the values and priorities the society holds. One key part of this field is focusing on the scale and location of important social **influencers**. This helps sociologists better predict how, say, a new idea or a new policy will affect us positively or negatively based on how quickly it reaches highly influential people in the network of social connections.

Think back to the picture in chapter 2 about finding shortest routes. Sociologists can use the same kind of diagram, but instead of filling it out with cities and the length of paths between them, they can build it to look at which "cities" (in this case, people) have the most roads connected to them. The people with more connections—especially to other people with a high number of connections—are considered the pivotal opinion leaders. Instead of trying to watch what everyone thinks or says in order to predict the society's reaction to a change, the sociologists show we can get a representative sample by focusing on listening to or influencing those high-value people.

As you might imagine, using the data in order to make predictions is important to the field—how long until the community becomes stable again after a military rebellion or a disease outbreak, for example. That means sociologists are going to rely on additional models like those from statistics and from graphing functions as central to their work.

Sociology has offered some surprising examples of ways that the models and tools it works with most closely turn out to have useful roles in other areas as well. One great example is to consider the interesting research showing that we can predict certain things about a company by imagining that it is a small country instead. When we do that, we can apply some of the sociology models to help understand and predict behaviors. As an example of this, just think about the way we could take the demographic models above and instead of asking the person's age, we could substitute in "how long the person has worked here." That would allow us to measure whether we are hiring enough new people and doing it when they could have time to learn from the long-timers before they retire and take their specialized knowledge with them.

Another interesting area where sociology models have turned out to be useful is in marketing. The same kind of dynamics that influence a community's reaction to a new president, for instance, also apply to a community's reaction to a new kind of technology being offered for sale. How the influencers react, how fast the use of the technology spreads to their closest network connections, and so on help a company understand the best way to market the new device. This information also helps make a prediction for how quickly the new device needs to be produced to keep up with how fast new groups react. Similarly, the marketers can watch for clues and cues that avoid creating a bad reaction among the influencers because that could also spread, and instead of responding like a new medicine, they might respond like it was a new plague.

Sports Coach

Athletic team coaches typically carry out some combination of the following and oversee support staff and assistant coaches who carry out much of the rest: analyze strengths and weaknesses of opposing players and teams, give guidance and direction in development and execution of game strategy, track and motivate the training and improvement of players, and run key elements of the business side, like making decisions about pay, player trades/drafts, and overseeing the pipeline for new talent.

In addition to building up the players, the coach is also responsible for helping to create a strong sense of team community and for serving as a representative to the public to help generate support for the team.

Here is a great example of something we talked about early on. Sports is the kind of career where you live and die by the averages—batting averages, earned run averages, average rushing yards per game, the average of judges' scores for gymnastics, and so on. The bread and butter of this area is going to be the bar chart which allows the coaches and managers to constantly make strategic decisions based on recent averages compared side by side.

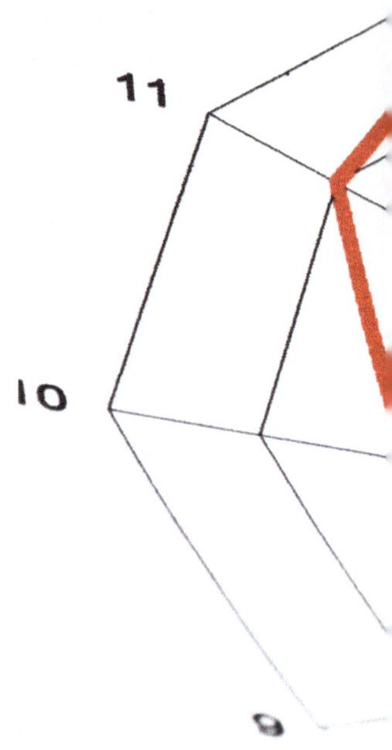

Player development

One of the two things I want to cover, though, is the radar chart. Sometimes it is also called a spider web chart (for obvious reasons). To the right is an example of what one might look like for the soccer team's two best goalies.

Each spoke on the wheel, numbered from 1 to 12, represents some measurement the coaching staff cares about. Number 1 could be, for instance, accuracy on penalty kicks. The coach might not care or spend a lot of time working on the goalies'

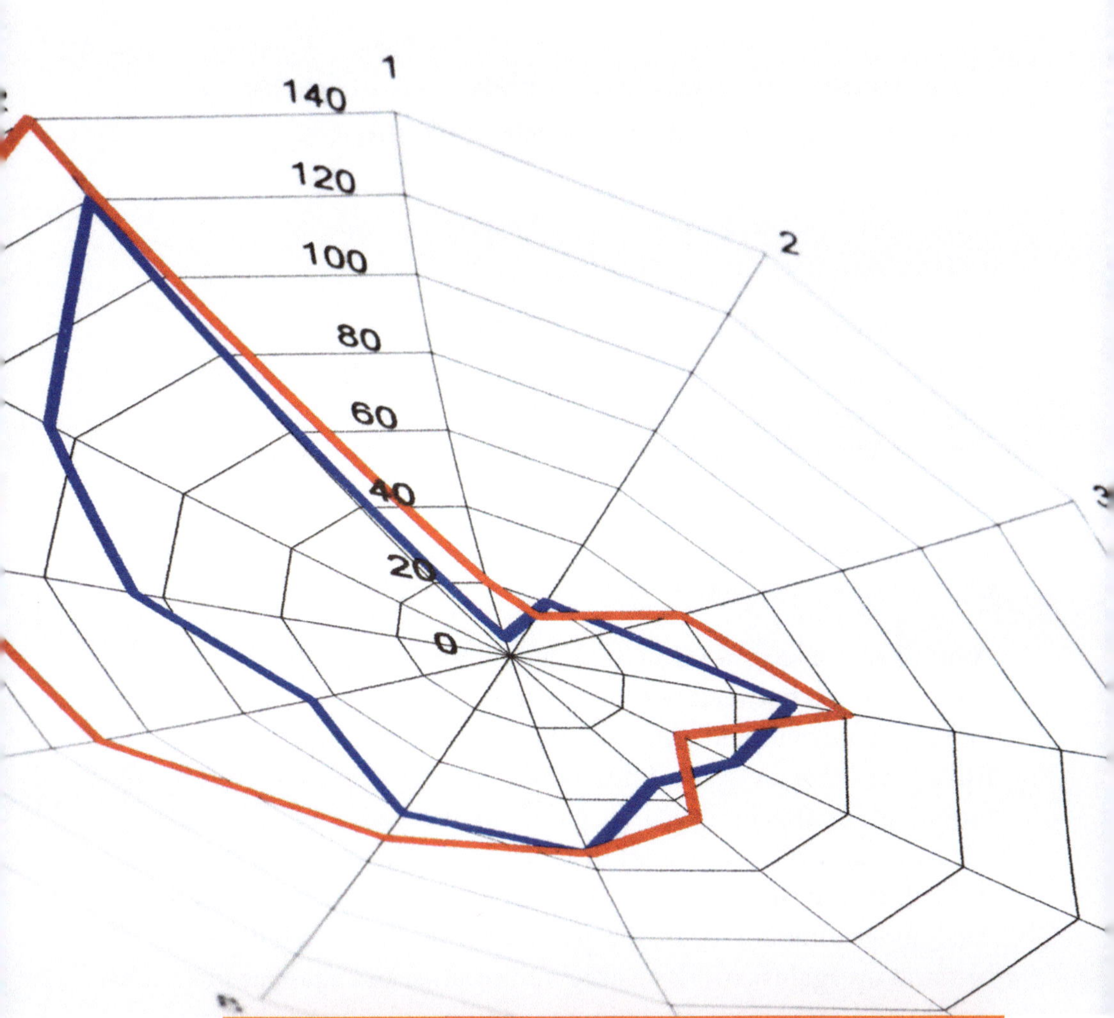

A radar chart like this allows us to easily see differences between two players on twelve different variables or skills.

Meet Tom Waechter

Please meet Tom Waechter, architect and Director of Capital Planning at the University of Kansas.

What kind of college degree did you get to prepare for being an architect and how long have you been at it?

My degree was a BS in architecture as part of a five-year program. I had the added advantage of two years of engineering school (lots of math) and graduate courses in real estate development (also more math, much more than you would first think).

Is there a kind of math or mathematical reasoning process that you find kind of interesting?

A lot of the process for designing a building continually shifts back and forth between working with the numbers and then with more subjective ingredients (like the look and feel of the design). I like that there are a lot of different kinds of math involved: the problems about geometry and space, rate problems, lighting and acoustics, and, of course, a lot of structural calculations having to do with technical performance—like the strength of materials used under different loads, the structural integrity of a building to stand up against wind, seismic movements, and changes in weight and movement of equipment and people. All of that goes into the mathematical modeling.

Is there an area of math that you wish you were better at? Why?

My two years of engineering were nearly all based on calculus and differential math, and I wish I'd been even a little further along in calculus. I'd like to understand the energy models behind building projects, as these are becoming more and more complex. It would also be nice to understand more about how to use math to understand the way large groups of equations interact with each other in different situations.

What are some cool new and creative ways math is being used in architecture?

Sustainability, energy use, and new materials that respond to the amount of sunlight to maintain adequate lighting or reduce heat loss: all of these are new components that get applied to many building projects today. And these all have a variety of associated math. Sometimes the math is built into the computer models, but because some architects don't understand the math, the ideas are not being applied as well as they could be.

How interesting. Thank you so much for sharing your perspective!

PKs, so it has a low, low score. Item number 12 might be reaction speed, which is crucial for a goalie. This chart gives the coach a very quick comparison in a way that several parts of a bar graph would not.

Note there are twelve variables being mapped in this case. It's possible for a radar chart to have more or fewer of them, as long as it was still relatively easy to use. (No one wants to try to sort one of these that measures one hundred variables.)

Moneyball

Many coaches also value scatterplots and finding the line of best fit (known as regression) among those dots. As many sports fans know, a few teams in recent years began having great success by using statistical models like the one on page 95 to test for which actions in the game have the most impact on winning. Based on what they found when they measured the different activities and how much correlation they had with winning, the more value they were given in the team's strategy going forward.

The basic idea is to think of it like this: Imagine a bunch of houses, represented by the dots, want to have a new road put in. Whether the houses are above the road or below it, the city wants to make sure to put the road in the place that is closest to the most houses, and that way it doesn't overpay for the water and sewer pipes to hook up to the main that will be under the street.

By measuring the slope of that new road, we can get a measurement that tells how closely connected the x and the y are. The more closely connected they are, the more we know that increasing the number of times we do something (x) will increase the number of times something else happens (y). For instance, the more times we get a runner on base, the more times we win.

The case of "moneyballing" (yes, it has become a verb) is of particular interest because what happened in that case is not the development of a new tool. Instead, analysts applied the same tools toward different, and more useful, variables. This strategy is called

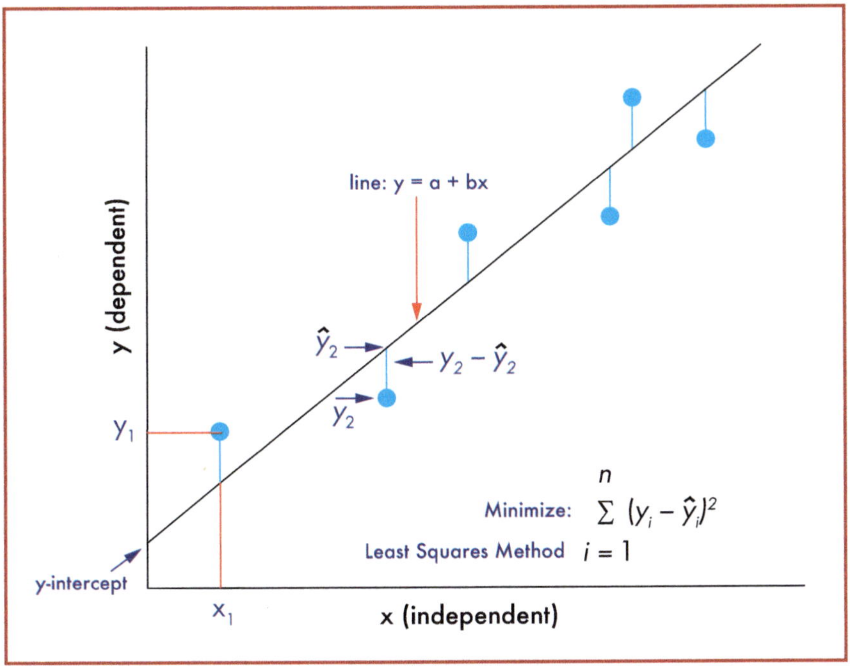

"moneyballing" because there was a famous book with the title *Moneyball* that told the story of the professional baseball team that first started using this strategy.

To illustrate the kind of research that caused a change, scouts and managers (and fans) used to think that numbers like a player's batting average and the number of homeruns he hit in a year were good statistics to measure the chances of scoring a lot of runs and winning a lot of games. Whichever teams had the most players with the highest batting averages should win the most pennants and world championships according to the old theory.

However, the statistics researchers discovered that hitting a lot of homeruns didn't actually tilt the balance of the game as much as things like how many times a player got on base. While other teams kept paying big money trying to compete against each other for the professional hitters with the best homerun averages, a team with a smaller budget could go in and get the players who managed

If a particle could travel through the black ring on the left side of this image and instantaneously arrive at the distant galaxy on the right side, what has it traveled through?

Wormholes

Albert Einstein and Nathan Rosen first proposed the idea of wormholes in 1935. Using the theory of **relativity**, they suggested two openings at different locations in space could exist so that going in one would allow you to come out the other.

It is important to point out that no one has discovered an actual wormhole yet. Those of you who learned about wormholes from science fiction books and movies will be disappointed to learn that the mathematical models only predict the existence of teeny-tiny little wormholes—about 4 to 13 inches (10 to 33 centimeters) across.

A second major problem is that the version predicted by Einstein's model would be likely to collapse quickly, so traveling through one would be impossible, even if we shrunk down to the normal size of a wormhole or we found one that had been stretched by the expansion of space.

Here's a challenge in understanding wormholes: The pictures I have seen show a wormhole like a tunnel connecting two different areas of space. But where is the tunnel? Think about this: Imagine your house is the universe. You climb out one window of your house and appear stepping in from another window. Now, there is nothing outside your house, since your house is *all* of space-time. How could there be an outside? Where are you between stepping out one window and stepping in another one?

to get on base the most. Since that was a more useful statistic for predicting wins, well, they got more wins!

Regression is a model that dampens the risk, both at the top and the bottom of the roller coaster. In business, as in being a stockbroker, the amount of ups and downs is referred to as volatility. That means this kind of approach can be valuable to apply to any field where there is room to make the behavior swings stay close to the prediction. Getting the situation to stay closer to the line of best fit improves predictability, which means, among other things, keeping costs down and being able to make smart decisions in planning strategy for the future.

Architects

Architects design and build offices, museums, factories, houses, and so on. Their job encompasses a wide range of tasks, from meeting with people to discuss the needs and ideas for a new building to putting together a bid for how much it would cost to design it and build it. They also supervise the construction once a plan has been decided on.

Besides those things, architects also coordinate and oversee a number of supporting specialists, from draftspeople and model builders to construction managers. They work with permits and zoning regulations along the way.

Turbulence simulations

There are many issues in architecture that we could use various models to represent, but let's focus on something that is much more uncommon: turbulence imaging.

The diagram on the next page shows the impact of building a new building on airflow and turbulence. An understanding of the different wind loads on the building will help to drive a number of decisions about frame structure, materials, and footprint to help make sure the building is stable over time. It can also affect the

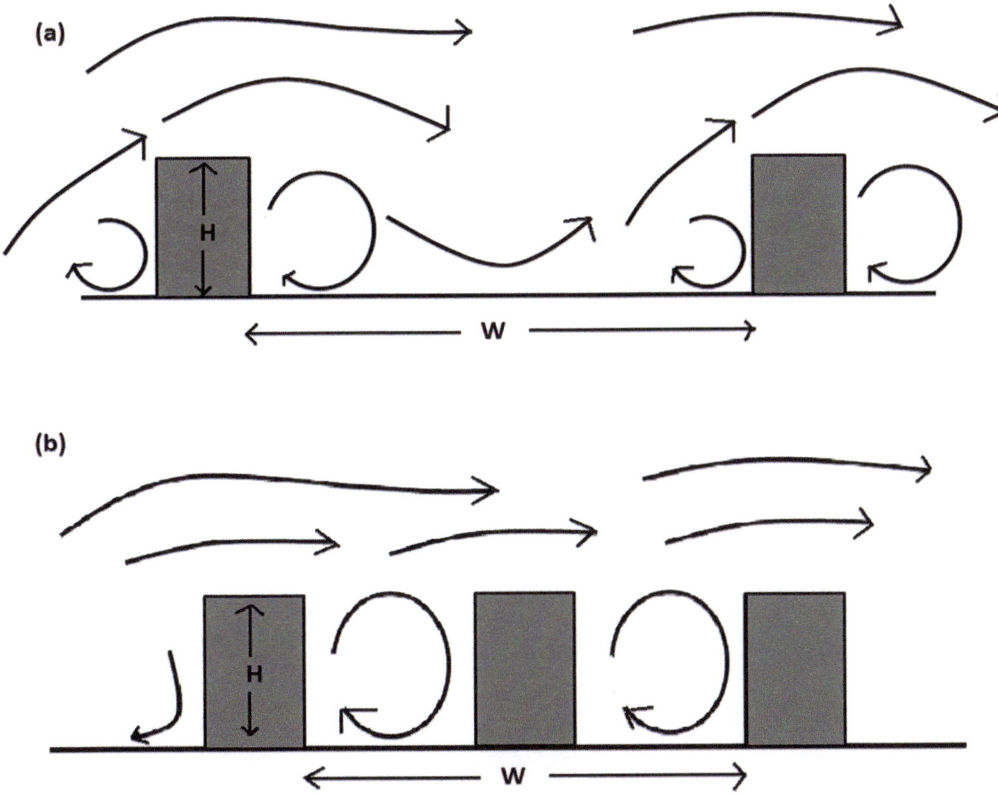

air exchange rates of the building (which need to be analyzed to keep the air inside the building clear and breathable) and can have an adverse impact on car and pedestrian traffic at the base of the building, so the architect needs to take the turbulence pattern into consideration.

Use the forces

Some examples of other key models that architects depend on are flow models, very similar to the transportation models talked about in chapter 2. These affect everything from air exchange rates to the patterns of foot traffic through the building. Additionally, architects will depend on vector diagrams to represent the

Francis Guthrie stumbled on one of the great mathematical puzzles in history—the four-color problem—that doesn't really need "normal math" to try to solve.

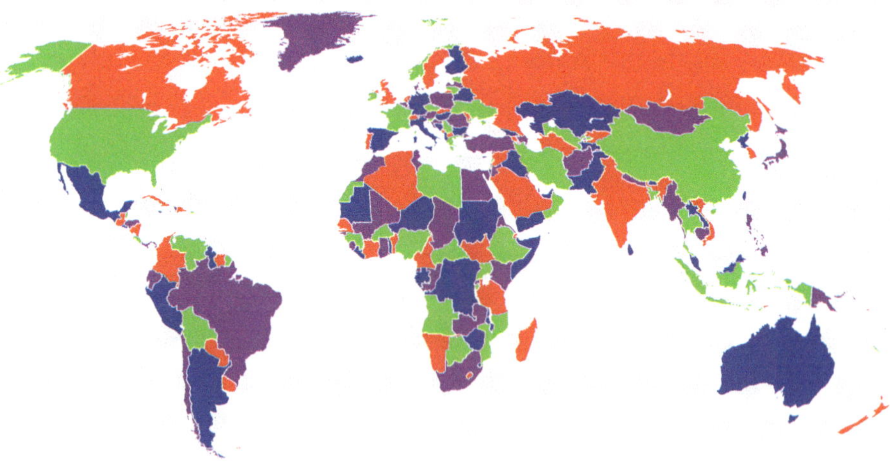

Four-Color Theorem

In 1852, an Englishman named Francis Guthrie was working on a map of the counties of England. As he was coloring them in, he noted that no matter how twisted the shapes, four colors seemed to be plenty so that no two shapes whose sides were touching would be the same color.

There have been a number of proofs attempting to solve this over the years. Some that seemed to succeed turned out to be disproven years later. Two proofs, one in 1879 by Alfred Kemp and a different one in 1880 by Peter Guthrie Tait, went for eleven years before being disproven. Currently, the proofs that exist rely on computers to test out certain parts of the proof, so many purists are not willing to concede a proof in this regard.

This is particularly important because there are so many test cases that have to be run in the solution, so there is no reasonable way to double-check the computer's proof (except by another computer). Therefore, we can only trust computers instead of holding the proof open until solved by a mathematician. Try playing with the idea on paper. Can you prove you only need four colors, regardless of the size and shape of the various regions on a map?

directions and amounts of different forces at play in the structure. They use **optimization** models to minimize materials and cost while still holding the project in line with the client and building code criteria.

Turbulence models can be slow to calculate and hard to communicate well with others who don't understand the complex mathematics, so these computer models go a long way to remove those two obstacles. They allow turbulence simulations to be used in areas ranging from aircraft design to animated movies which need to imitate the behavior of water or smoke, both of which exhibit complex turbulence. Even within architecture, we see a variety of other uses for these models. Examples include managing the air exchange rates and the effect of water movement on bridge supports that are anchored underwater. Turbulence models also play a significant role in our circulatory system.

Meteorologist

Meteorologists are weather scientists who study weather patterns, climate, and trends impacting humans. Some of their roles include developing forecasts, assisting in the design and testing of new models or tools for analyzing collected data, and advising their clients or the viewing public on upcoming weather events.

Among their many duties, meteorologists collect and analyze rain, wind, and temperature data (collected from a variety of sources, including weather balloons, radar systems, and satellites) and plug it all into complex math and computer models. Once they run those models, they are in a position to give long-term and short-term forecasts about the weather.

Remember that not all meteorologists are on TV. Some of them are hired on to provide longer-range forecasting, policy advising, and identification of shifting trends in weather conditions (in rainfall, for instance, which affects water supply).

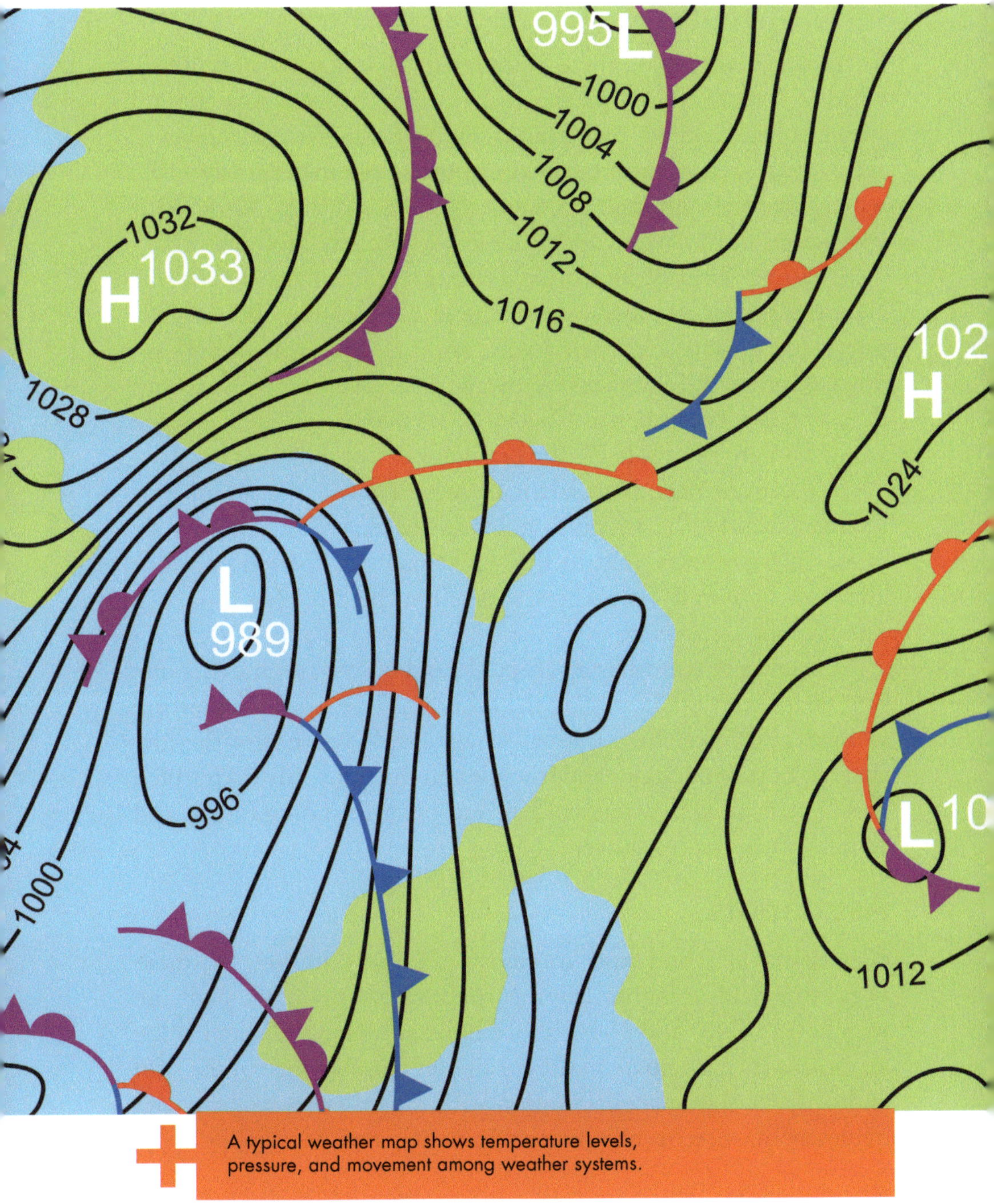

A typical weather map shows temperature levels, pressure, and movement among weather systems.

Fronts and pressures

There is actually much more complex math going on in building weather forecasts than people realize, and it would take several whole college courses to explain how it works. We can, however, see a small slice of that reflected in the sample map on page 103, which shows fronts and pressures for a given area, along with indications of relative temperature levels compared to each other.

The map kind of looks like a topographical map, but in this case, the higher numbers mean higher air pressure. The highest points of pressure are labeled with an *H*. Those areas of air sink down. Generally, high pressure means clear skies and light winds. Areas of low pressure are labeled *L*. Low-pressure areas can be windy and are favorable for the development of rain or snow.

We can also see blue triangles, which mean the air in that section is cold—the triangles point in the direction the cold air is moving. Compare that to the red semicircles that show warm air fronts and point (you know, sort of) in the direction those big lumps of air are moving.

The areas of the different "fronts" (cold body of air moving one way, warm body of air moving the other way) are where interesting weather stuff, like storms, usually happens. If you notice the line that has both semicircles and triangles on the same side (in purple), that signals an area where cold air has pushed in underneath and trapped warm air up above.

Astronauts

Astronauts are, first and foremost, scientists in one or more respective fields, ranging from astronomy to botany to medicine to robotics. Over and above those normal career skills and roles, all astronauts participate in the repair and maintenance of the ship and/or space station, collaboration with the other researchers, and deployment or retrieval of any equipment, parts, or supplies.

Astronauts go through rigorous training for thorough knowledge of the different systems of the ship/station. On top of that, they will all have been briefed on mission requirements or objectives.

There are not particularly unique models I wanted to show you in this section, as the astronauts would be using a variety of other models we have already seen from other areas. What I wanted you think about is whether we should count dials, gauges, and light arrays as models. Do they represent the real world through mathematics?

Ask yourself why converting a temperature in the engine to a number and having it appear on a computer graph is a model, but having that temperature turn into a needle pointing to a certain section of a gauge isn't. What is different?

To the vector go the spoils

Aside from the reliance on gauges, dials, and displays, there are some other models with which astronauts would need to be perfectly comfortable, such as vector analysis for the movement of a ship, deployed equipment, and extravehicular excursions. They would also need to be familiar with a small pile of linear function curves for rates of consumption (fuel, food, water, oxygen, time, battery charges, and more).

Geographer

Geographers research and analyze different variables related to the physical, human, and socio-cultural landscape. They look at things like concentrations of people and/or cultural groups and where they are located, distances between high-density settlement areas, changing sizes or compositions of different communities, and so on.

Some of the different parts of their jobs include gathering data from observations in the real world, analyzing satellite maps, and conducting interviews and independent research. The other key

This view of a space shuttle cockpit shows many gauges, lights, and displays which model each element of the complex machine.

part of their job is to use the information and research they've gathered in order to create useful maps of different variables as they shift over time and in relation to each other.

Comparing models of population density

Below is a simple bar graph depicting the population of major American cities. It contains some interesting and useful information, and it certainly allows us to compare the cities to each other. But overall, the data is fairly limited.

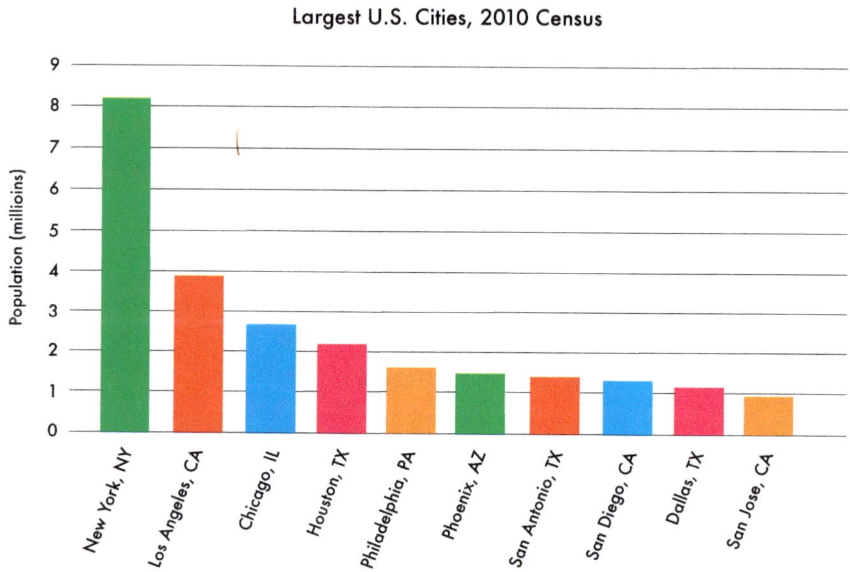

Compare the graph with the map on page 109 which allows you to see the spatial relationship between the areas of different densities:

See how there is an additional layer of information in this case that allows us to zero in on possible explanations and causes for such wide-ranging gaps in the population density? Populations in the United States are higher in the East, where European settlement began. Populations also tend to be higher near majors sources of water, such as oceans, the Great Lakes, and rivers.

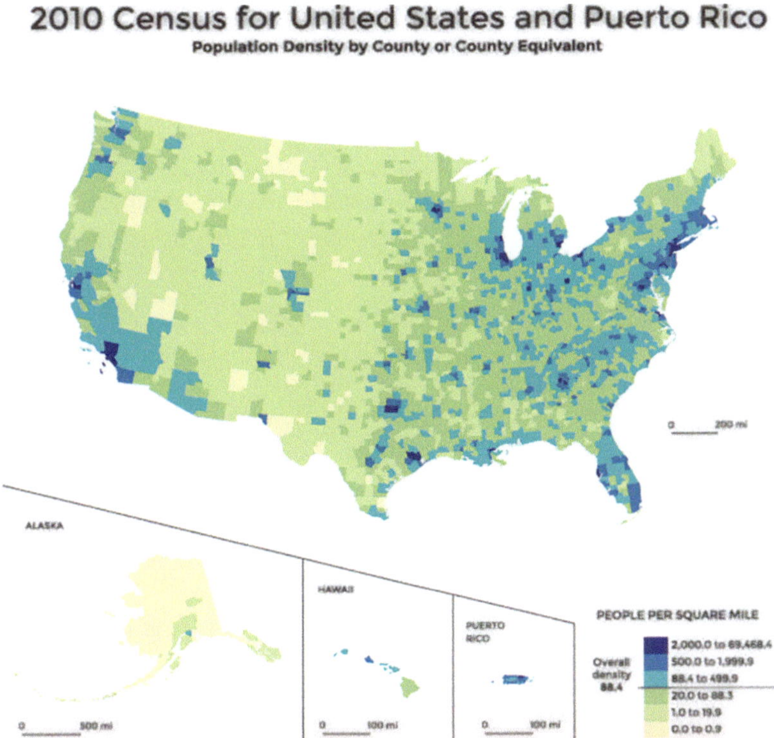

In this map of the United States, different colors show differences in population density for different areas.

Not all maps and crayons

Geography is much more than map-making. Geography is about the relationship and impact between humans and the natural landscape. Some geographers work in the field, interviewing people to understand attitudes and migration. They then put the collected field data together with other variables, like how many people immigrated or emigrated over the last year. They then run regression analysis to check for correlations with portrayals of a given region in the news or in political speeches. The idea is that the human construct of "The Middle East" or "The Great Plains" has an impact on other attitudes and feelings about living there.

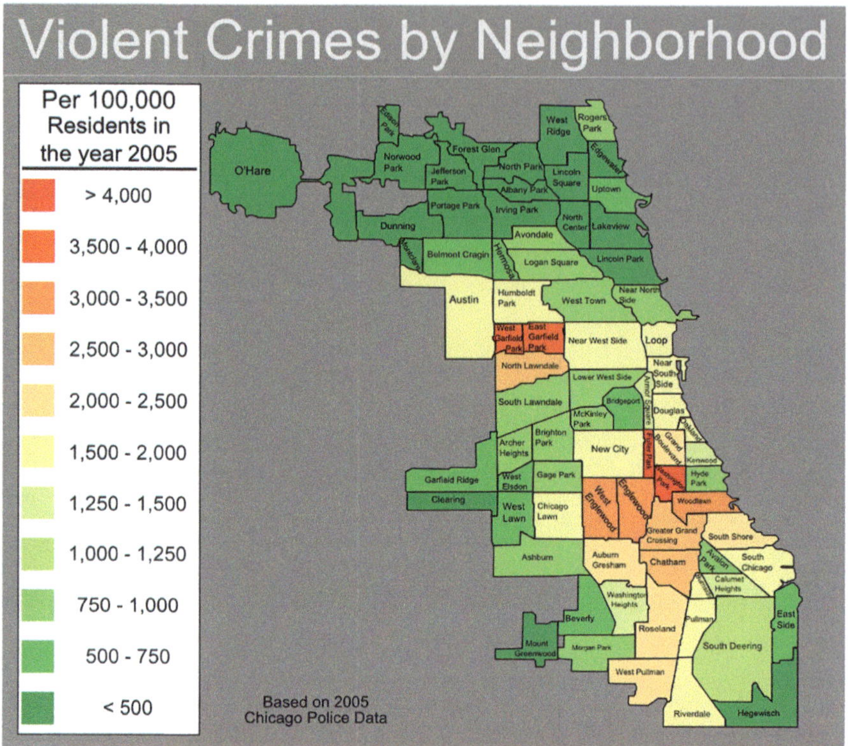

This map uses color to indicate how many crimes were committed in different areas of Chicago, Illinois, in 2005.

DEA Agent

Drug Enforcement Agency (DEA) agents are a special subset of federal law enforcement tasked with narrow focus on enforcing the laws relating to illegal drugs. As law enforcement officers, they share many common job characteristics with detectives and police on patrol. They respond to incidents, pursue and arrest suspects, gather evidence from crime scenes, and document their activities. They also work with people and organizations in the community to educate the public and to strengthen various drug prevention and awareness programs. However, unlike local law enforcement officers, DEA agents have a wider jurisdiction in order to pursue those who have broken federal drug laws across the boundaries that would otherwise limit normal police officers.

Geographic profiling

One of the tools that agents use to solve a case is a geographical map for the areas where incidents of crime occurred. This data helps the agents to look for possible suspects and leads in particular areas with which the criminal seems familiar or related in some way. For example, these might be areas near where the criminal works or lives.

Data mining

An additional key model in the DEA approach is a kind of route analysis in order to search out the main pathways by which drugs might be coming into a given area. Agents also use lots and lots of regression to look for ways that behavior clues and evidence from a particular case can be found to match up with certain characteristics of the perpetrator. Those might match to physical characteristics or personality traits.

Management Consultant

The career of management consulting is to serve as a strategic advisor to other companies and management teams. Consultants' services usually cover a wide range of different subject areas including finance, marketing, IT, organizational development, and strategic planning.

A company will usually hire a management consulting firm to guide and assist with major projects for which they do not normally have enough expertise already on board or for which they need an objective outside team. Often the consulting firm is brought in to help select and oversee a major change of direction by the company in one or more areas. Examples might be cutting some products to focus on a narrower specialization, changing the way the company is organized, or converting to a new strategy and business model to help the company adapt to changes in the marketplace.

Consultants usually work as part of a five-person or six-person teams for a given client project. Beyond specific client projects, consultants assist with building general resource libraries for the consulting firm, industry research, and networking to help build relationships with other potential client companies.

Charts and Models, Inc.

These people are the usually the top 1 percent of the top 1 percent at what they do. They have to be—otherwise why would major corporations hire them and pay them incredible amounts of money to come in and help solve complicated problems that the company wasn't able to solve on its own? Consulting is commonly one of the highest-paying options around. (Though in fairness, the amount of time spent traveling to client companies, away from family and friends, and the workload are also in the top 1 percent of the top 1 percent.)

Consultants don't just need to come in and figure out a solution to the big, messy, expensive problem which could cause the client company to implode and lose a lot of its market share if handled badly. They also need to be able to communicate the solutions clearly to the client and create a set of tools and plans that will help the company navigate the transition to the new way of doing things. After consultants come in, do their research, and crunch the numbers, the pitch for the new plan and the launch are centered around a big set of charts, diagrams, and models—about 200 or 250 pages of them, not counting the reports and detailed analysis that resulted in the numbers and percentages shown on those charts.

The point of this final example, then, is not to illustrate a particular new or different kind of model that can be used for

One of the most valuable uses of models is allowing us to find patterns that will allow us to make good predictions about a business's growth.

getting a better understanding of certain key numbers and trends in a given business. The point is to help by suggesting that this is about building parts of a map to help get you wherever you want to go.

So, we've wandered around for a while among a variety of different career fields and seen some of their models and explanations. Hopefully you've gradually acquired a better appreciation for the value these tools add to anyone who wants to use math to help them pursue success in their chosen field. Models can help describe the universe at the tiniest level, and they can juggle the complex diplomatic elements of helping your country pursue success at the conference table with other nations.

CONCLUSION

Hey, you made it! The last person coming through the museum exhibit got lost and the security guards found him curled up under a bench by the theoretical physicists display, so well done. What did you think about the ideas and examples?

At the beginning of the book, I talked briefly about how math equations are themselves models, but perhaps too far trimmed down to provide us with our best grasp of how they relate to things going on in our normal experience.

To come around and connect the pieces, I would like to leave you with a slightly more mind-blowing idea. It is ancient in origins, with thinkers like Plato and Pythagoras of ancient Greece offering some of the earliest written versions of the idea.

The idea is that mathematics is actually the truest picture of the universe, not just some set of symbols we use to talk about it. Just think along for a minute. If we wanted to talk about, say, the sun, we couldn't very well claim that mathematics was its true nature, right? I mean, that would sound like it didn't have a nature or could not have been described by a race of aliens who might have encountered it a million years before we came along. (I don't mean there have to be aliens. This is a thought experiment.) The sun still had the same chemical structure before any race came along and gave those chemicals names. The sun had the same explosive personality, the

Our world is not swirling through empty space. Instead, we are swimming through a sea of equations.

same problem with sunspots that always seem to show up on the day of school pictures, and so on.

So what are those properties and those chemicals if we take away the name tags we stuck on them? They are sets of particles with different numbers (number of neutrons, mass, etc.).What is more, theoretical physicists think that even those particles are made up of nearly invisible "strings" that are characterized just by their frequency number.

Along this line of thinking, then, just allow that the universe itself might be made of math, in which case the math equations and formulas are the most accurate picture of it, not the "least accurate" in kind of the way we talked about it before. Brilliant thinkers like Galileo and Kepler would agree. They thought that math was the secret code in which the universe was written.

Take that with you as you turn this final page to close the book and set it back on its shelf. What if math *is* the secret language of the universe? And what if models are a way to translate that language into something we can understand in the real world?

Glossary

aerobic Exercise intended to improve the health of respiratory and circulatory systems.

assessments Tests administered to measure current skill sets in some specific area.

assumptions Items taken as true without the intent or the ability to prove them.

Buddhist An adherent of Buddhism, a religion originating in ancient India following the teachings of Buddha (whose birth name was Siddhartha Gautama).

comparative Numbers or measures that are intended for use in comparing two or more things.

component A part or element of a larger whole.

continuous Forming a smooth, unbroken series of items, categories, or data.

cyclical Behaving in a regular, repeating series.

data Facts or statistics collected for the primary purpose of research.

demographics Statistical data relating to a population or subgroups within it.

economics The branch of study of producing, trading, and consuming goods and services.

electrodes Conductors used to send or receive electrical signals from a non-metallic part of a circuit.

equilibrium The condition when two opposing forces or influences are balanced so as to cancel each other out.

escalate To become or cause to become higher or more intense.

frequency The rate at which something occurs (repeats) in a given period of time.

function An equation that assigns one output value (a *y*) to each input value (an *x*).

hypothesis A claim supported by reason that can be tested by experimentation.

influencers Individuals who have the power and persuasion to affect the attitudes or buying behavior of others.

interval A repeated break, gap, or space between two or more things.

magnitude The degree or extent of something in reference to an initial value.

optimization The process of finding the highest or lowest possible value for a given function.

ornate Having complex decorations or intricate details.

pictogram A pictorial symbol that stands for a word or phrase.

portfolio A collection of sample work. Also a collection of stocks and other investments.

proportion A relationship of equality between two ratios.

quarks Any of a number of subatomic particles believed to make up atoms.

refraction The bending of rays of light or sound as they move from one medium to another.

regression Using math to analyze a set of data points to come up with an equation for the line that is closest to the data points overall.

relativity The theory in physics explaining how certain features and properties depend on the situation of the observer.

scenario A particular setting or series of events.

social science Refers to any of the areas of research involving systems of action and interaction involving human beliefs and behaviors—for example, history or sociology.

standardized tests Tests in which everyone answers the same questions and statistical analysis is employed to make sure they are evaluated in a consistent way across large groups.

strategy A plan or method intended toward the achievement of some particular goal.

substantial Of sufficient size or worth as to make a noticeable impact on the results.

supplemental Something that is extra or added on.

translated To describe or express something in another language or visual system.

turbulence When fluids (including air and water) are being redirected within the stream by an object in their path, interfering with the flow.

vague Unclear or ambiguous.

Further Reading

Books

Gowers, Timothy, June Barrow-Green, and Imre Leader, eds. *The Princeton Companion to Mathematics*. Princeton, NJ: Princeton University Press, 2008.

Huettenmueller, Rhonda. *Precalculus Demystified: Hard Stuff Made Easy*. New York: McGraw-Hill, 2012.

Jackson, Tom, ed. *Mathematics: An Illustrated History of Numbers*. New York: Shelter Harbor Press, 2012.

Pickover, Clifford A. *The Math Book: From Pythagoras to the 57th Dimension, 250 Milestones in the History of Mathematics*. New York: Sterling, 2009.

Simmons, George F. *Precalculus Mathematics in a Nutshell: Geometry, Algebra, Trigonometry*. Eugene, OR: Wipf & Stock, 2003.

Websites

Cool Math

www.coolmath.com/precalculus-review-calculus-intro

This is a good website for a wide range of topics. Of particular note is the colorful layout for a good basic review of precalculus material.

Drexel University Math Forum

mathforum.org/dr.math/

This site offers a variety of basic explanations of concepts and examples. One of its strengths is that the topics are broken out by subjects and by level in school (elementary, middle, high school, college and beyond). They do a helpful job of working through example problems and also allow you to search by topic.

Get the Math

www.thirteen.org/get-the-math/

This website has some good videos geared at the middle-school level that give interaction with real-life examples from fashion, music, etc.

Khan Academy

www.khanacademy.org

This is an excellent online learning website, and the learn-as-you-go format allows you to click on only the short instructional videos you need as you work your way through the course. There are courses at a wide range of skill levels.

NYU Department of Mathematics—Courant Institute

cims.nyu.edu/~kiryl/precalculus.html

This is a nice, clean, visual layout of a thorough variety of topics in trigonometry and other transcendental function topics, as well as review of key ideas from algebra.

Online Chart Tool

www.onlinecharttool.com

This site has a nice collection of sample graphs and diagrams you can play around with to see how they work.

***Plus* Magazine**

plus.maths.org/content/

This online magazine is run under the Millennium Mathematics Project at Cambridge University. It offers interesting short articles on interesting problems in math (and science). The problems are complex, but they are broken down in a way that is easy to follow.

Bibliography

Barnett, Raymond A, Michael R. Ziegler, and Karl E. Byleen. *College Algebra with Trigonometry*. New York: McGraw-Hill, 2001.

Bellos, Alex. *The Grapes of Math*. New York: Simon and Schuster, 2014.

Berlinghoff, William P., Kerry E. Grant, and Dale Skrien. A *Mathematical Sampler: Topics for Liberal Arts*. Lanham, MD: Ardsley House Publishers, 2001.

Byers, William. *How Mathematicians Think: Using Ambiguity, Contradiction, and Paradox to Create Mathematics*. Princeton, NJ: Princeton University Press, 2007.

Charles, Randall I., *Algebra I: Common Core*. New York: Pearson, 2010.

Clawson, Calvin C. *Mathematical Mysteries: The Beauty and Magic of Numbers*. New York: Plenum Press, 1996.

Devlin, Keith. *Mathematics: The Science of Patterns: The Search for Order in Life, Mind and the Universe*. New York: Henry Holt, 2003.

Ellenberg, Jordan. *How Not to Be Wrong: The Power of Mathematical Thinking*. New York: Penguin, 2014.

Freeman, W. H., Brendan Cady, ed. *For All Practical Purposes: Mathematical Literacy in Today's World*. New York: W. H. Freeman and Company, 2006.

Frenkel, Edward. *Love & Math: The Heart of Hidden Reality*. New York: Basic Books, 2013.

Gordon, John N., Ralph V. McGrew, and Raymond A. Serway. *Physics for Scientists and Engineers*. Belmont: Thomson Brooks/Cole, 2005.

Jackson, Tom, ed. *Mathematics: An Illustrated History of Numbers*. New York: Shelter Harbor Press, 2012.

Lehrman, Robert L. *Physics The Easy Way*. Hauppage, NY: Barron's Educational Series, Inc., 1990.

Mankiewicz, Richard. *The Story of Mathematics*. Princeton, NJ: Princeton University Press, 2000.

McLeish, John. *Numbers: The History of Numbers and How They Shape Our Lives*. New York: Fawcett Columbine, 1991.

Oakley, Barber. *A Mind for Numbers: How to Excel at Math and Science (Even if You Flunked Algebra)*. New York: Tarcher, 2014

Paulos, John Allen. *Beyond Numeracy: Ruminations of a Numbers Man*. New York: Alfred A. Knopf, 1991.

Peterson, Ivars. *Islands of Truth: A Mathematical Mystery Cruise*. New York: W. H. Freeman and Company, 1990.

—. *The Mathematical Tourist: Snapshots of Modern Mathematics*. New York: W. H. Freeman and Company, 1988.

Pickover, Clifford A. *The Math Book: From Pythagoras to the 57th Dimension, 250 Milestones in the History of Mathematics*. New York: Sterling, 2009.

Rooney, Anne. *The Story of Mathematics: From Creating the Pyramids to Exploring Infinity*. London: Arcturus, 2015.

Sardar, Ziauddin, Jerry Ravetz, and Borin Van Loon. *Introducing Mathematics*. Cambridge, UK: Icon Books, 1999.

Stewart, Ian. *Nature's Numbers: The Unreal Reality of Mathematics*. New York: Basic Books, 1995.

Stewart, James, Lothar Redlin, and Saleem Watson. *Elementary Functions*. Boston: Cengage Learning, 2011.

Strogatz, Steven. *The Joy of X: A Guided Tour of Math, from One to Infinity*. Boston: Houghton-Mifflin, 2012.

Index

Page numbers in **boldface** are illustrations. Entries in **boldface** are glossary terms.

About the Author

Erik Richardson is an award-winning teacher from Milwaukee, where he has taught and tutored math up to the college level over the last ten years. He has done graduate work in math, economics, and the philosophy of math, and he uses all three in his work as a business consultant with corporations and small businesses. He is a member of the Kappa Mu Epsilon math honor society, and some of his work applying math to different kinds of problems has shown up at conferences, in magazines, and even in a few pieces of published poetry. As the director of Every Einstein (everyeinstein.org), he works actively to get math and science resources into the hands of teachers and students all over the country.